Selected Titles in This Series

692 **Mara D. Neusel,** Inverse invariant theory and Steenrod operations, 2000
691 **Bruce Hughes and Stratos Prassidis,** Control and relaxation over the circle, 2000
690 **Robert Rumely, Chi Fong Lau, and Robert Varley,** Existence of the sectional capacity, 2000
689 **M. A. Dickmann and F. Miraglia,** Special groups: Boolean-theoretic methods in the theory of quadratic forms, 2000
688 **Piotr Hajłasz and Pekka Koskela,** Sobolev met Poincaré, 2000
687 **Guy David and Stephen Semmes,** Uniform rectifiability and quasiminimizing sets of arbitrary codimension, 2000
686 **L. Gaunce Lewis, Jr.,** Splitting theorems for certain equivariant spectra, 2000
685 **Jean-Luc Joly, Guy Metivier, and Jeffrey Rauch,** Caustics for dissipative semilinear oscillations, 2000
684 **Harvey I. Blau, Bangteng Xu, Z. Arad, E. Fisman, V. Miloslavsky, and M. Muzychuk,** Homogeneous integral table algebras of degree three: A trilogy, 2000
683 **Serge Bouc,** Non-additive exact functors and tensor induction for Mackey functors, 2000
682 **Martin Majewski,** ational homotopical models and uniqueness, 2000
681 **David P. Blecher, Paul S. Muhly, and Vern I. Paulsen,** Categories of operator modules (Morita equivalence and projective modules, 2000
680 **Joachim Zacharias,** Continuous tensor products and Arveson's spectral C^*-algebras, 2000
679 **Y. A. Abramovich and A. K. Kitover,** Inverses of disjointness preserving operators, 2000
678 **Wilhelm Stannat,** The theory of generalized Dirichlet forms and its applications in analysis and stochastics, 1999
677 **Volodymyr V. Lyubashenko,** Squared Hopf algebras, 1999
676 **S. Strelitz,** Asymptotics for solutions of linear differential equations having turning points with applications, 1999
675 **Michael B. Marcus and Jay Rosen,** Renormalized self-intersection local times and Wick power chaos processes, 1999
674 **R. Lawther and D. M. Testerman,** A_1 subgroups of exceptional algebraic groups, 1999
673 **John Lott,** Diffeomorphisms and noncommutative analytic torsion, 1999
672 **Yael Karshon,** Periodic Hamiltonian flows on four dimensional manifolds, 1999
671 **Andrzej Rosłanowski and Saharon Shelah,** Norms on possibilities I: Forcing with trees and creatures, 1999
670 **Steve Jackson,** A computation of δ^1_5, 1999
669 **Seán Keel and James M^c^Kernan,** Rational curves on quasi-projective surfaces, 1999
668 **E. N. Dancer and P. Poláčik,** Realization of vector fields and dynamics of spatially homogeneous parabolic equations, 1999
667 **Ethan Akin,** Simplicial dynamical systems, 1999
666 **Mark Hovey and Neil P. Strickland,** Morava K-theories and localisation, 1999
665 **George Lawrence Ashline,** The defect relation of meromorphic maps on parabolic manifolds, 1999
664 **Xia Chen,** Limit theorems for functionals of ergodic Markov chains with general state space, 1999
663 **Ola Bratteli and Palle E. T. Jorgensen,** Iterated function systems and permutation representation of the Cuntz algebra, 1999
662 **B. H. Bowditch,** Treelike structures arising from continua and convergence groups, 1999
661 **J. P. C. Greenlees,** Rational S^1-equivariant stable homotopy theory, 1999
660 **Dale E. Alspach,** Tensor products and independent sums of $\mathcal{L}_p$-spaces, $1 < p < \infty$, 1999

(Continued in the back of this publication)

Inverse Invariant Theory and Steenrod Operations

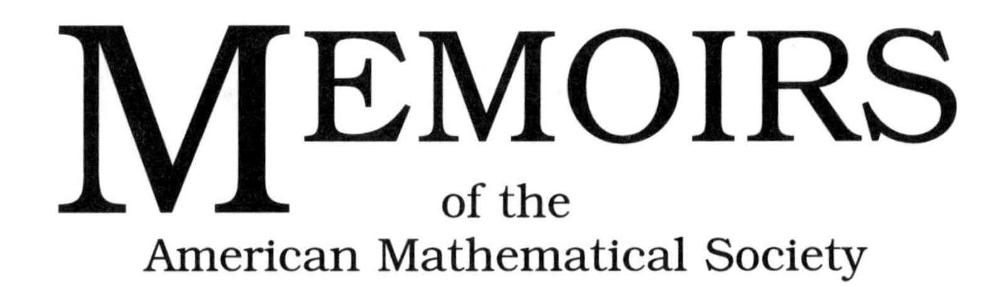

Number 692

Inverse Invariant Theory and Steenrod Operations

Mara D. Neusel

July 2000 • Volume 146 • Number 692 (first of 5 numbers) • ISSN 0065-9266

American Mathematical Society
Providence, Rhode Island

2000 *Mathematics Subject Classification.*
Primary 13A50, 55S10; Secondary 55–XX, 13–XX.

Library of Congress Cataloging-in-Publication Data

Neusel, Mara D., 1964–
Inverse invariant theory and Steenrod operations / Mara D. Neusel.
p. cm. — (Memoirs of the American Mathematical Society, ISSN 0065-9266 ; no. 692)
Includes bibliographical references.
ISBN 0-8218-2091-5 (alk. paper)
1. Steenrod algebra. 2. Invariants. I. Title. II. Series.
QA3 .A57 no. 692
[QA612.782]
510 s—dc21
[512.55] 00-036255

Memoirs of the American Mathematical Society

This journal is devoted entirely to research in pure and applied mathematics.

Subscription information. The 2000 subscription begins with volume 143 and consists of six mailings, each containing one or more numbers. Subscription prices for 2000 are $466 list, $419 institutional member. A late charge of 10% of the subscription price will be imposed on orders received from nonmembers after January 1 of the subscription year. Subscribers outside the United States and India must pay a postage surcharge of $30; subscribers in India must pay a postage surcharge of $43. Expedited delivery to destinations in North America $35; elsewhere $130. Each number may be ordered separately; *please specify number* when ordering an individual number. For prices and titles of recently released numbers, see the New Publications sections of the *Notices of the American Mathematical Society.*

Back number information. For back issues see the *AMS Catalog of Publications.*

Subscriptions and orders should be addressed to the American Mathematical Society, P. O. Box 5904, Boston, MA 02206-5904. *All orders must be accompanied by payment.* Other correspondence should be addressed to Box 6248, Providence, RI 02940-6248.

Memoirs of the American Mathematical Society is published bimonthly (each volume consisting usually of more than one number) by the American Mathematical Society at 201 Charles Street, Providence, RI 02904-2294. Periodicals postage paid at Providence, RI. Postmaster: Send address changes to Memoirs, American Mathematical Society, P. O. Box 6248, Providence, RI 02940-6248.

This publication is indexed in *Science Citation Index*®, *SciSearch*®, *Research Alert*®, *CompuMath Citation Index*®, *Current Contents*®*/Physical, Chemical & Earth Sciences.*
Printed in the United States of America.

∞ The paper used in this book is acid-free and falls within the guidelines established to ensure permanence and durability.
Visit the AMS home page at URL: `http://www.ams.org/`

10 9 8 7 6 5 4 3 2 1 05 04 03 02 01 00

Contents

ABSTRACT : This paper is devoted to the study of inverse invariant theory and its relationship with the $\mathcal{P}^*$–invariant prime spectrum of an unstable algebra over the Steenrod algebra. We will show that this spectrum is a chain saturated poset, which inherits a number of classical properties from the full prime spectrum. In any unstable algebra we prove the existence of Thom classes, which we use to find a fractal of the Dickson algebra, and give a counterexample to the Reverse Landweber–Stong Conjecture. Along the way to these results we will generalize the important theorems of Adams–Wilkerson to arbitrary Galois fields. To do so we examine closely fields and their extensions over the Steenrod algebra, and generalize some results about the unstable part of a module over the Steenrod algebra. In addition we construct the inseparable closure of an unstable algebra over the Steenrod algebra, investigate its properties and calculate important examples.

ACKNOWLEDGEMENT : This research was done essentially during my visit to Queen's University, Kingston Ontario, Canada, during the fall term 1997. This visit was partially supported by the NSERC of Canada. I want to thank H. E. A. Campbell for the invitation to Kingston, and for many comments and suggestions on a preliminary version of this paper, and the members of the invariant theory seminar, the curves seminar and the Hughes family for their hospitality. The last few bells and whistles were completed at the University of Minnesota, Minneapolis, in the spring of 1998 while I was a visiting assistant professor. A version of this manuscript was submitted to the Georg-August Universität, Göttingen, in August 1998 as *Habilitationsschrift*.

I thank Leslie G. Roberts for his gorgeous examples and Ljudmila Bordag, Ram Murty and Marie A. Vitulli for their assistance during winter 1997/8.

I am deeply indebted to Larry Smith for proof reading, correcting and commenting many versions of this paper as well as for many fruitful discussions.

I thank Kathrin Kuhnigk for her boundless comments on the manuscript, Jeanne Duflot for a series of critical letters, which put a finger on many weak points, and the referee for her/his careful corrections of some stupidities and many helpful suggestions and comments. Their efforts to make this manuscript agreeable to read are much appreciated.

Received by the editor June 17, 1998, revised May 5, 1999.

Introduction

Yaşamak bir ağaç gibi tek ve hür
ve bir orman gibi kardeşçesine,
bu hasret bizim ...
Nazım Hikmet

Let $\mathbb{F} := \mathbb{F}_q$ be a Galois field of characteristic p and with $q = p^s$ elements. If

$$\rho : G \hookrightarrow \mathrm{GL}(n,\ \mathbb{F})$$

is a faithful representation of a finite group G of degree n over the field $\mathbb{F}$, then ρ induces an action of the group G on the polynomial algebra

$$\mathbb{F}[V] = \mathbb{F}[x_1, \ldots, x_n]$$

in n linear indeterminates. The ring of polynomials invariant of this action, denoted by $\mathbb{F}[V]^G$, is then a graded connected[1] commutative Noetherian algebra over $\mathbb{F}$ of Krull dimension n; see [36] for an introduction to the invariant theory of finite groups.

Invariant theory of finite groups investigates the ring of polynomial invariants, $\mathbb{F}[V]^G$, for, e.g., a given group G or a family of representations; its homological, algebraic and combinatorial properties. The classical inverse question is:

Let H^* ***be a ring. When does a representation exist such that***
$$\mathrm{H}^* = \mathbb{F}[V]^G?$$

Let's start by listing some significant properties of any ring of invariants, $\mathbb{F}[V]^G$. If H^* is to be a ring of invariants it too must have these properties.

[1] The terminology **connected** comes from algebraic topology. It means that in degree zero the algebra consists of the ground field only, i.e.,

$$\left(\mathbb{F}[V]^G\right)_{(0)} = \mathbb{F}.$$

PROPERTY 1: By what we said so far we certainly need to assume that H^* is a graded connected commutative Noetherian algebra over $\mathbb{F}$. Denote by $\dim(\mathrm{H}^*) = n$ the Krull dimension of H^*. Let me first say some more words about the **grading convention**: It is encoded in the notation

$$\mathrm{H}^* = \left\{ (\mathrm{H}^*)_{(i)} \mid i \in \mathbb{N}_0 \right\},$$

where $\mathbb{N}_0$ denotes the non-negative integers, i.e., using the terminology introduced by J. W. Milnor and J. C. Moore, H^* is a **non-negatively graded algebra**. Since sign conventions play no role in this manuscript we prefer to use this standard grading convention of commutative algebra. If you are a topologist, you may in certain situations, want to double the degrees to bring things into conformity with the standard grading conventions of algebraic topology.

PROPERTY 2: Secondly, we need that H^* embeds integrally[2] into the polynomial ring $\mathbb{F}[V]$, since, by a classical result of Emmy Noether,

$$\mathbb{F}[V]^G \hookrightarrow \mathbb{F}[V]$$

is an integral extension for every finite group G.

PROPERTY 3: Thirdly, the full general linear group $\mathrm{GL}(n, \mathbb{F})$ is a finite group (because our ground field is finite) and provides us with a set of universal invariants, i.e.,

$$\mathcal{D}^*(n) = \mathbb{F}[\mathbf{d}_{n,0}, \ldots, \mathbf{d}_{n,n-1}] = \mathbb{F}[V]^{\mathrm{GL}(n, \mathbb{F})} \hookrightarrow \mathbb{F}[V]^G$$

is an integral extension for any G. The ring of polynomial invariants of $\mathrm{GL}(n, \mathbb{F})$ was calculated by L. E. Dickson in 1911, [9], and is therefore called **Dickson algebra**, compare also [36] §8.1. The generating polynomials

$$\mathbf{d}_{n,0}, \ldots, \mathbf{d}_{n,n-1}$$

are called **Dickson classes**. (One might look up their explicit description in [36] Section 8.1 or in the Appendix A.2.) So, since $G \hookrightarrow \mathrm{GL}(n, \mathbb{F})$ implies $\mathbb{F}[V]^{\mathrm{GL}(n, \mathbb{F})} \hookrightarrow \mathbb{F}[V]^G$, we need to assume that there exists an integral extension

$$\mathcal{D}^*(n) \hookrightarrow \mathrm{H}^*.$$

PROPERTY 4: A fourth property of rings of invariants is a classical result of Emmy Noether: they are integrally closed (i.e., in their field of fractions), see [36] Proposition 1.2.4.

OBSERVATION : *Let H^* be a graded connected integrally closed commutative Noetherian algebra over $\mathbb{F}$. Suppose there are integral extensions*

$$\mathcal{D}^*(n) \hookrightarrow \mathrm{H}^* \hookrightarrow \mathbb{F}[V].$$

Then H^* is a ring of invariants.

---

[2] For two rings A^* and B^*, here and in the following "A^* embeds integrally into B^*" or "B^* contains integrally A^*" is meant to be a short way of saying that the ring extension $A^* \hookrightarrow B^*$ is integral.

PROOF: By what we have assumed we have the following situation

$$\begin{array}{ccccc} \mathcal{D}^*(n) & \underset{\text{integral}}{\hookrightarrow} & \mathrm{H}^* = \overline{\mathrm{H}^*} & \underset{\text{integral}}{\hookrightarrow} & \mathbb{F}[V] \\ \downarrow & & \downarrow & & \downarrow \\ FF(\mathcal{D}^*(n)) & \hookrightarrow & FF(\mathrm{H}^*) & \hookrightarrow & \mathbb{F}(V), \end{array}$$

where $FF(-)$ denotes the field of fractions functor and $\overline{\mathrm{H}^*}$ the integral closure of H^*. The field extension $FF(\mathcal{D}^*(n)) \subseteq \mathbb{F}(V)$ is Galois with Galois group $\mathrm{GL}(n, \mathbb{F})$. The Fundamental Theorem of Galois Theory gives us a group $G \leq \mathrm{GL}(n, \mathbb{F})$ such that

$$FF(\mathrm{H}^*) = \mathbb{F}(V)^G \subseteq \mathbb{F}(V).$$

Hence we have an integral extension

$$\mathrm{H}^* \hookrightarrow \mathbb{F}[V]^G$$

of integrally closed algebras with the same field of fractions $FF(\mathrm{H}^*) = \mathbb{F}(V)^G$, so they are equal

$$\mathrm{H}^* = \mathbb{F}[V]^G.$$

That's all we claimed •

So, in order to solve the inverse invariant theory problem we have to characterize those H^*'s which:

PROPERTY 2: admit an integral embedding into $\mathbb{F}[V]$, and

PROPERTY 3: contain integrally $\mathcal{D}^*(n)$.

However, note carefully, that in the above proof it would have been enough to assume that we had a commutative diagram

$$\begin{array}{ccccc} & & \mathrm{H}^* & \underset{\text{integral}}{\hookrightarrow} & \mathbb{F}[V] \\ & & \downarrow & & \downarrow \\ \mathbb{F}(V)^{\mathrm{GL}(n,\, \mathbb{F})} = FF(\mathcal{D}^*(n)) & \hookrightarrow & FF(\mathrm{H}^*) & \hookrightarrow & \mathbb{F}(V) \end{array}$$

such that the field extension

$$\mathbb{F}(V)/FF(\mathcal{D}^*(n))$$

is Galois. So, we will characterize those H^*'s which fit into such a diagram, i.e., which

PROPERTY 2: admit an integral embedding into $\mathbb{F}[V]$, and

PROPERTY 3': whose field of fractions contains the field of fractions of the Dickson algebra

$$FF(\mathcal{D}^*(n)) \hookrightarrow FF(\mathrm{H}^*) \hookrightarrow \mathbb{F}(V)$$

such that the extension $FF(\mathcal{D}^*(n)) \subseteq \mathbb{F}(V)$ is a Galois extension with Galois group $\mathrm{GL}(n, \mathbb{F})$.

At this point you have to recall that we are working over a finite field. This fact gives us an additional structure: the **Steenrod algebra** $\mathcal{P}^*$.

Denote[3] by $\mathbb{F}[V][[\xi]]$ the power series ring over $\mathbb{F}[V]$ in an additional variable ξ, and set $\deg(\xi) = 1 - q$. Define an $\mathbb{F}$-algebra homomorphism of degree[4] zero

$$\boldsymbol{P}(\xi) : \mathbb{F}[V] \longrightarrow \mathbb{F}[V][[\xi]],$$

by requiring

$$\boldsymbol{P}(\xi)(l) := l + l^q \xi \in \mathbb{F}[V][[\xi]] \quad \forall \text{ linear forms } l \in V^*.$$

In this way, we get, for an arbitrary polynomial $f \in \mathbb{F}[V]$, by separating out homogeneous components

$$(\star) \qquad \boldsymbol{P}(\xi)(f) = \begin{cases} \sum_{i=0}^{\infty} \mathcal{P}^i(f)\xi^i & \text{if } p \text{ is odd} \\ \sum_{i=0}^{\infty} \mathrm{Sq}^i(f)\xi^i & \text{if } p = 2, \end{cases}$$

where $\mathcal{P}^i$, resp. Sq^i, are certain $\mathbb{F}$-linear maps

$$\mathcal{P}^i, \ \mathrm{Sq}^i : \mathbb{F}[V] \longrightarrow \mathbb{F}[V].$$

These maps are functorial in V. The operations $\mathcal{P}^i$, resp. Sq^i are called **Steenrod reduced power operations**, resp. **Steenrod squaring operations**, or collectively **Steenrod operations** for short. In order to avoid double notation for the case $p = 2$, with the indulgence of topologists, we set $\mathrm{Sq}^i := \mathcal{P}^i$ for all $i \in \mathbb{N}_0$.

The sums appearing in $(\star)$ are actually finite. In fact $\boldsymbol{P}(\xi)(f)$ is a *polynomial* in ξ of degree $\deg(f)$ with leading coefficient f^q. This means the Steenrod operations satisfy the **unstability condition**

$$\mathcal{P}^i(f) = \begin{cases} f^q & \text{if } i = \deg(f) \\ 0 & \text{if } i > \deg(f) \end{cases} \quad \forall f \in \mathbb{F}[V].$$

Next, observe that the multiplicativity of the operator $\boldsymbol{P}(\xi)$ leads to the **Cartan formulae** for the Steenrod operations:

$$\mathcal{P}^k(f'f'') = \sum_{i+j=k} \mathcal{P}^i(f')\mathcal{P}^j(f'') \quad \forall f',\ f'' \in \mathbb{F}[V].$$

---

[3] The following description of the Steenrod algebra is due to Larry Smith, Sections 10.3 and 11.1 in [36]. Its big advantages are: first of all it is *algebraic*, i.e., avoids the whole issue of cohomology operations, and secondly is a uniform description for all Galois fields, i.e., *not only for the prime field*.

[4] Note carefully we ignore the usual topological sign conventions, since graded commutation rules play no role here.

So far we have just a bunch of operations acting in a certain way on $\mathbb{F}[V]$. We next pack all these Steenrod operations $\mathcal{P}^0 = \mathrm{id}$, $\mathcal{P}^1$, $\mathcal{P}^2$, ... in an $\mathbb{F}$-algebra, i.e., we form the $\mathbb{F}$-subalgebra of the endomorphism algebra $\mathrm{End}(\mathbb{F}[V])$ of the functor

$$\mathbb{F}[-] : \mathrm{Vect}_{\mathbb{F}} \longrightarrow \mathrm{GCA}_{\mathbb{F}}$$

from $\mathbb{F}$-vector spaces to graded commutative $\mathbb{F}$-algebras generated by these.

DEFINITION: The **Steenrod algebra** $\mathcal{P}^*$ is the $\mathbb{F}$-subalgebra of $\mathrm{End}(\mathbb{F}[-])$ generated by $\mathcal{P}^0$, $\mathcal{P}^1$, $\mathcal{P}^2$, ...

The generators satisfy certain commutation rules, namely the **Adem-Wu relations**:

$$\mathcal{P}^i \mathcal{P}^j = \sum_{k=0}^{[i/q]} (-1)^{i-qk} \binom{(q-1)(j-k)-1}{i-qk} \mathcal{P}^{i+j-k} \mathcal{P}^k \quad \forall\ i,\ j \geq 0,\ i < qj.$$

Note that the coefficients are elements in the prime field $\mathbb{F}_p$ of $\mathbb{F}$. A proof of these relations, can be found in Sections 10.3 and 11.1 in [36], see also, of course, the originals [7] and [47]. The **Bullett-Macdonald identity** provides us with a well-wrapped description of the relations among the Steenrod operations, [7]. To wit: set $u := (1-t)^{q-1} = 1 + t + \cdots + t^{q-1}$ and $s := tu$, then

$$\boldsymbol{P}(s) \circ \boldsymbol{P}(1) = \boldsymbol{P}(u) \circ \boldsymbol{P}(t^q).$$

Since $\boldsymbol{P}(\xi)$ is additive and multiplicative, it is enough to check this equation for the basis elements of V^*, which is indeed a short calculation.

By theorems of Henri Cartan, [8], Jean-Pierre Serre, [32], and Wu Wen Tsün, [47], see Chapter 11 in [36] for the case of arbitrary Galois fields, one has

THEOREM : *The Steenrod algebra $\mathcal{P}^*$ is the $\mathbb{F}$-algebra generated by the reduced power operations $\mathcal{P}^0$, $\mathcal{P}^1$, $\mathcal{P}^2$, ... modulo the Adem-Wu relations, resp. Bullett-Macdonald identity.*

The Bullett-Macdonald identity is compatible with the diagonal map given by the Cartan formulae

$$\begin{array}{ccc} \mathcal{P}^* & \longrightarrow & \mathcal{P}^* \otimes \mathcal{P}^* \\ \mathcal{P}^k & \longmapsto & \sum_{i+j=k} \mathcal{P}^i \otimes \mathcal{P}^j. \end{array}$$

The comultiplication so defined gives the Steenrod algebra the structure of a Hopf algebra, a result which was originally[5] proven by John W. Milnor for the case of prime fields, Theorem 1 in [25], using a mixture of topological and algebraic methods.

Well, we have a Hopf algebra. So, what's the dual? This question was also answered in the beautiful paper of Milnor, Theorem 2 in [25] - again over prime fields. However, since the coefficients of the Adem-Wu relations always lie in the prime field, Milnor's proof can be copied word-by-word and we get:

---

[5] For yet another approach and a proof over arbitrary Galois fields see Jeanne Duflot's Example 2.0.11 in [11].

THEOREM: *The dual algebra $\mathcal{P}_*$ of the Steenrod algebra $\mathcal{P}^*$ is a polynomial algebra generated by infinitely many elements $\xi_1, \xi_2, \ldots$, which are dual to the successively defined derivations[6] in the Steenrod algebra $\mathcal{P}^*$*

$$\mathcal{P}^{\Delta_i} := \begin{cases} \mathcal{P}^1 & \text{if } i = 1 \\ \mathcal{P}^{\Delta_{i-1}}\mathcal{P}^{q^{i-1}} - \mathcal{P}^{q^{i-1}}\mathcal{P}^{\Delta_{i-1}} & \text{otherwise.} \end{cases}$$

To summarize: the Steenrod algebra is just a particular way to encode information which is hidden in the classical **Frobenius homomorphism**. Nevertheless, the Frobenius map itself is *not* an element of the Steenrod algebra.

DEFINITION: A graded connected commutative algebra H^* over $\mathbb{F}$ is called **an algebra over the Steenrod algebra** if it is a left $\mathcal{P}^*$-module satisfying the Cartan formulae. If, in addition, the unstability condition holds then we say H^* is an **unstable algebra over the Steenrod algebra**, or simply **an unstable algebra**.

So, our polynomial algebra $\mathbb{F}[V]$ is by construction an unstable algebra over the Steenrod algebra. Coming back to our inverse invariant theory problem we notice that the action of the group G on $\mathbb{F}[V]$ commutes with raising to q-th powers. Hence every ring of invariants is an unstable[7] algebra over the Steenrod algebra $\mathcal{P}^*$. We therefore work throughout the whole manuscript in the **category whose objects are algebras over the Steenrod algebra $\mathcal{P}^*$ and whose morphisms are $\mathcal{P}^*$-module homomorphisms**.

So, our H^* in question, will be throughout the entire manuscript a graded, connected, commutative, $\mathbb{F}$-algebra over the Steenrod algebra $\mathcal{P}^*$, and all maps will commute with the action of the Steenrod algebra.

Equipped with this terminology we can describe the results and organization of this manuscript.

The first goal is to show that our H^* embeds integrally into $\mathbb{F}[V]$ if and only if H^* is a Noetherian integral domain. The one statement is trivial, the other, is for prime fields $\mathbb{F} = \mathbb{F}_p$, the contents of the famous Adams-Wilkerson Embedding Theorem, [1], see also Section 25 in [18]. So, we are aiming at a generalization of this theorem to arbitrary Galois fields.

To this end we need some technical preliminaries, which we will collect, following Larry Smith, in the Δ-Theorem, compare Section 5, in particular Theorem 5.1, in [1] or [36] Theorem 10.5.4. We will introduce notions of **Δ-finiteness** and **Δ-length**, which will be central in later chapters. Moreover, we will round things out with some examples, and draw some consequences. This is the contents of Chapter 1.

---

[6] These derivations play a crucial role in everything what follows. We will have in Section 1.1 a more intimate encounter with them.

[7] For a natural example of an algebra over $\mathcal{P}^*$, which is not unstable, see Chapter 4.

In Chapter 2 we will provide a proper setup for the category of graded fields over the Steenrod algebra and generalize C. W. Wilkerson's Separable Extension Lemma, [43], or Lemma 10.5.3 in [36]. Moreover, we will introduce the notion of **purely inseparable field extensions** *in this category* and **inseparably closed fields over $\mathcal{P}^*$**. We will develop some of their properties and illustrate them with examples.

In Chapter 3 we will have a closer look at the unstable part of a **module** over $\mathcal{P}^*$ (n.b. here only the second half of the unstability condition makes sense). We generalize some results from [44], in particular the Integral Closure Theorem, see also Theorem 10.5.2 in [36].

In Chapter 4 we will define the **$\mathcal{P}^*$-inseparable closure** $\sqrt[\mathcal{P}^*]{\mathrm{H}^*}$ of an algebra H^* over the Steenrod algebra, give a construction method, develop its main properties, and give illustrative examples. It will turn out, that, under appropriate assumptions, taking the $\mathcal{P}^*$-inseparable closure commutes with taking the field of fractions.

This will allow us to prove the Embedding Theorem for $\mathcal{P}^*$-inseparably closed unstable algebras H^* in Chapter 5. We will start this chapter with the investigation of **PROPERTY 3'**. It turns out that for unstable integral domains H^*, we can single out in $FF(\mathrm{H}^*)$ a (possibly trivial) $FF(\mathcal{D}^*(m))$, where the transcendence degree m of $FF(\mathcal{D}^*(m))$ over $\mathbb{F}$ is precisely the Δ-length of H^* as defined in Chapter 1. However, the field extension

$$FF(\mathcal{D}^*(m)) \hookrightarrow FF(\mathrm{H}^*)$$

might not be algebraic. We will characterize exactly when it is. This leads in Section 5.2 to Theorem 5.2.1, one of the main technical ingredients (compare [36] Theorem 10.5.5) for the Embedding Theorem, which will be proved in Section 5.3.

The first important result of Chapter 6 is Theorem 6.1.3, where we prove that for an unstable integral domain H^*, $\sqrt[\mathcal{P}^*]{\mathrm{H}^*}$ is Noetherian if and only if H^* is Noetherian. This allows us to complete the proof of the Embedding Theorem in full generality, Corollary 6.1.5. After that we will have a rest and enjoy deriving some nice results about the $\mathcal{P}^*$-invariant homogeneous prime spectrum, $\mathcal{P}roj_{\mathcal{P}^*}(\mathrm{H}^*)$, of an unstable Noetherian H^* in Section 6.2, which all follow quite naturally. This investigation was one motivation that started all this. We will prove that $\mathcal{P}roj_{\mathcal{P}^*}(\mathrm{H}^*)$ forms a chain saturated subposet of the full spectrum of homogeneous prime ideals, in particular:

(1) $\mathcal{P}roj_{\mathcal{P}^*}(\mathrm{H}^*)$ is a finite set,
(2) for any $i = 0, \ldots, n = \dim(\mathrm{H}^*)$ there is a prime ideal $\mathfrak{p}_i \in \mathcal{P}roj_{\mathcal{P}^*}(\mathrm{H}^*)$ of height i,
(3) for any $\mathfrak{p} \in \mathcal{P}roj_{\mathcal{P}^*}(\mathrm{H}^*)$ there exists a saturated ascending chain of prime ideals

$$\mathfrak{p} = \mathfrak{p}_i \subsetneq \mathfrak{p}_{i+1} \subsetneq \cdots \subsetneq \mathfrak{p}_n = \mathfrak{m} \subset \mathrm{H}^*$$

starting at $\mathfrak{p}$ and ending at the maximal ideal $\mathfrak{m}$ of H^*, all of which are contained in $\mathcal{P}roj_{\mathcal{P}^*}(\mathrm{H}^*)$, and

(4) for any $\mathfrak{p} \in \mathcal{P}roj_{\mathcal{P}^*}(\mathrm{H}^*)$ there exists a saturated descending chain of prime ideals

$$\mathfrak{p}_0 \subsetneqq \mathfrak{p}_1 \subsetneqq \cdots \subsetneqq \mathfrak{p}_i = \mathfrak{p} \subset \mathrm{H}^*$$

starting at a minimal prime ideal $\mathfrak{p}_0$ of H^* and ending at $\mathfrak{p}$, all of which are contained in $\mathcal{P}roj_{\mathcal{P}^*}(\mathrm{H}^*)$.

In Section 6.3 we will also show that for an unstable reduced algebra H^*, i.e., $\mathcal{N}il(\mathrm{H}^*) = (0)$, the $\mathcal{P}^*$-inseparable closure $\sqrt[\mathcal{P}^*]{\mathrm{H}^*}$ is Noetherian if and only H^* is Noetherian.

It then takes just a tiny bit more to prove the Galois Embedding Theorem for arbitrary Galois fields, [1], see also Section 25 in [18]:

THEOREM (Galois Embedding Theorem): *Let* H^* *be an unstable algebra over* $\mathbb{F}$. *Then* H^* *is a ring of invariants* $\mathbb{F}[V]^G$ *if and only if* H^* *is an integrally closed,* $\mathcal{P}^*$*-inseparably closed, Noetherian integral domain.*

This we will prove in Chapter 7. We will also show what we can do if we had a diagram like

$$\begin{array}{ccc} \mathrm{H}^* & \underset{\text{integral}}{\hookrightarrow} & \mathbb{F}[V] \\ \downarrow & & \downarrow \\ FF(\mathrm{H}^*) & \underset{\text{inseparable}}{\hookrightarrow} & \mathbb{F}(V). \end{array}$$

This will generalize results by [46]. Moreover, we will show that the existence of a Dickson algebra integrally inside H^* (**PROPERTY 3**) is determined by the existence of enough p-th roots in H^*. If this is not the case we can only find a fractal of the Dickson algebra[8] in H^*. This is the Little Imbedding Theorem.

Note carefully that **PROPERTY 2**, namely that

$$\mathrm{H}^* \underset{\text{integral}}{\hookrightarrow} \mathbb{F}[V]$$

reflects faithfully the non-existence of zero divisors, while **PROPERTY 3**

$$\mathcal{D}^*(n) \underset{\text{integral}}{\hookrightarrow} \mathrm{H}^*$$

expresses the existence of enough p-th roots. This last statement remains true for unstable algebras *with* zero divisors.

---

[8] A **fractal of the Dickson algebra** $\mathcal{D}^*(n)$ is a q-th power

$$\begin{aligned}\mathcal{D}^*(n)^{q^l} &= \mathbb{F}[\mathbf{d}_{n,0}^{q^l}, \ldots, \mathbf{d}_{n,n-1}^{q^l}] \\ &= \mathbb{F}[x_1^{q^l}, \ldots, x_n^{q^l}]^{\mathrm{GL}(n,\,\mathbb{F})},\end{aligned}$$

where one can look up the verification of the last equality in [36] in the proof of Theorem 11.4.6.

The final Chapter 8 contains additional work on unstable algebras with zero divisors and some results about $\mathcal{P}roj_{\mathcal{P}^*}(\mathrm{H}^*)$. We prove that even if H^* has zero divisors it is an integral extension of a fractal of the Dickson algebra. This is the Big Imbedding Theorem. Finally, we prove the existence of **Thom classes**, i.e., elements which generate a $\mathcal{P}^*$-invariant principle ideal of height[9] one. In addition to this we construct a Thom class $\mathbf{t} \in \mathrm{H}^*$ which is contained in every $\mathcal{P}^*$-invariant homogeneous prime ideal of positive height. From this we are able to draw additional conclusions about the $\mathcal{P}^*$-invariant homogeneous prime spectrum. We will show that there is a $\mathcal{P}^*$-invariant version of Krull's Principal Ideal Theorem and its generalization, i.e., for any $\mathcal{P}^*$-invariant prime ideal $\mathfrak{p} \subset \mathrm{H}^*$ of height $i \in \{1, \ldots, n = \dim(\mathrm{H}^*)\}$ there exist i elements $h_1, \ldots, h_i \in \mathfrak{p}$ such that

(1) $\mathfrak{p}$ is an isolated prime ideal of $(h_1, \ldots, h_i) \subset \mathrm{H}^*$,
(2) the ideals $(h_1, \ldots, h_j) \subset \mathrm{H}^*$ for $j = 1, \ldots, i$ are $\mathcal{P}^*$-invariant of height j.

These proofs exemplify the utility of Thom classes: they allow proofs by induction over the Krull dimension. In the last section we will look at the **Reverse Landweber-Stong Conjecture**. We need some terminology: A sequence $h_1, \ldots, h_k \in \mathrm{H}^*$ of elements of positive degree in H^* is called a **regular sequence** if

(1) $h_1 \in \mathrm{H}^*$ is not a zero divisor,
(2) $h_i \in \mathrm{H}^*/(h_1, \ldots, h_{i-1})$ is not a zero divisor $\forall\ i = 2, \ldots, k$.

Then define the **depth** (or **homological codimension**) of H^*, denoted $dp(\mathrm{H}^*)$, to be the length of the longest possible regular sequence in H^*. See [36] Chapter 6 for an introduction to the homological properties of graded connected commutative Noetherian $\mathbb{F}$-algebras.

The original **Landweber-Stong Conjecture** asserts that a ring of invariants $\mathbb{F}[V]^G$ has depth at least k, $dp(\mathrm{H}^*) \geq k$, if and only if the k *bottom* Dickson classes[10] $\mathbf{d}_{n,n-1}, \ldots, \mathbf{d}_{n,n-k}$ form a regular sequence, see [22]. This conjecture was proven in 1996 by Dorra Bourguiba and Saïd Zarati, [6], using the classification of injective $\mathbb{F}[V]$-modules over $\mathcal{P}^*$ by J. Lannes and S. Zarati, see [23]. For a slightly different approach see the survey article by Larry Smith, [37].

In the aftermath Larry Smith, [38], asked whether the **Reverse Landweber-Stong Conjecture** is true, i.e., whether a ring of invariants $\mathbb{F}[V]^G$ has depth at least k if and only if the k *top* Dickson classes, $\mathbf{d}_{n,0}, \ldots, \mathbf{d}_{n,k-1}$ form a regular

[9] Recall that the height of an arbitrary ideal $I \subset R$ in a commutative ring R with unity, is defined to be the minimal $k \in \mathbb{N}_0$ such that there exists a prime ideal $\mathfrak{p}$ of height k containing I, i.e.,

$$ht(I) := \min\{k \in \mathbb{N}_0 \mid \exists\ \mathfrak{p} \text{ prime}\,,\ ht(\mathfrak{p}) = k,\ I \subseteq \mathfrak{p}\}.$$

[10] Note that since we have a ring of invariants, of course, the Dickson algebra $\mathcal{D}^*(n) \hookrightarrow \mathbb{F}[V]^G$ is integrally contained in the ring of invariants.

sequence.[11] Since the Big Imbedding Theorem hands us a fractal of the Dickson algebra in any unstable algebra over the Steenrod algebra, we can formulate the Reverse Landweber-Stong Conjecture more generally as follows:

REVERSE LANDWEBER-STONG CONJECTURE: Let H^* be an unstable algebra over the Steenrod algebra. Then H^* has depth at least k if and only if high enough q-th powers of the k top Dickson classes, $\mathbf{d}_{n,0}^{q^l}, \ldots, \mathbf{d}_{n,k-1}^{q^l} \in H^*$ form a regular sequence.

We will give a counterexample to this conjecture. This emphasizes the significance of the original Landweber-Stong Conjecture and answers one of the many questions about the structure of algebras over the Steenrod algebra, which were the original motivation to start this investigation.

I exiled some technical stuff into the appendix.

I hope you gonna enjoy the reading of this paper and the nice results presented here.

[11] When we take into account the $\mathcal{P}^*$-invariant homogeneous prime spectrum of H^*, and in particular of the Dickson algebra $\mathcal{D}^*(n)$, then this version looks a lot more natural than the original Landweber-Stong Conjecture.

Chapter 1
The Δ-Theorem

In this chapter we prepare ourselves for the journey we plan to make, i.e., we will provide the tools we will need in later chapters. In particular, we will prove the Δ-Theorem (see the parents [1] Theorem 5.1 or the godfather [36] Theorem 10.5.4) in its most general form.

Throughout the whole chapter H^* is an unstable algebra over the Steenrod algebra.

1.1 Steenrod Operations

We need a bit terminology to be able to explain what we are about to do. Recall the following definition:

DEFINITION: *A map $D : \mathrm{H}^* \longrightarrow \mathrm{H}^*$ is called a* **derivation**, *if*

(1) *D acts linearly on H^*, and*

(2) *$D(hh') = D(h)h' + hD(h')$ for all $h,\ h' \in \mathrm{H}^*$.*

The inductively defined elements of the Steenrod algebra

$$\begin{aligned} \mathcal{P}^{\Delta_1} &:= \mathcal{P}^1, \\ \mathcal{P}^{\Delta_i} &:= [\mathcal{P}^{\Delta_{i-1}}, \mathcal{P}^{q^{i-1}}],\ i \geq 2, \end{aligned}$$

are primitive derivations[1], see[2], e.g., [25] Corollary 5 in §6. Note in addition that

$$\mathcal{P}^{\Delta_0}(h) := \deg(h)h,\ \forall\ h \in \mathrm{H}^*$$

---

[1] An element c in a Hopf algebra is called **primitive** if its image under the comultiplication is $c \otimes 1 + 1 \otimes c$.

[2] These elements are the Q^i's in the notation of J. F. Adams and C. W. Wilkerson, resp. the $P^{(0,\ldots,0,\,1,\,0,\ldots)}$'s (with the 1 in the i-th position), in J. W. Milnor's.

defines a derivation (that is not in $\mathcal{P}^*$). Observe that[3]

$$\deg(\mathcal{P}^{\Delta_i}) = q^i - 1, \ i \geq 0.$$

REMARK: The derivations defined above satisfy the following commutation rules

$$[\mathcal{P}^{\Delta_i}, \ \mathcal{P}^{\Delta_j}] = \begin{cases} \mathcal{P}^{\Delta_i} & \text{for } i \neq 0 \text{ and } j = 0 \\ -\mathcal{P}^{\Delta_j} & \text{for } i = 0 \text{ and } j \neq 0 \\ 0 & \text{otherwise} \end{cases}$$

for all $i, \ j \geq 0$, compare Lemma 10.6.4 in [36] or apply Theorem 4b of [25].

OBSERVATION 1.1.1: *The set of all derivations $\mathcal{P}^{\Delta_i}$, $i \geq 0$, generates a left H^*-submodule of the endomorphisms of H^*, $\mathrm{End}(\mathrm{H}^*)$, which we will denote by $\Delta(\mathrm{H}^*)$, since, for every $a_0, \ldots, a_n \in \mathrm{H}^*$, and every $i_0, \ldots, i_n \geq 0$*

$$\delta := a_0\mathcal{P}^{\Delta_{i_0}} + \cdots + a_n\mathcal{P}^{\Delta_{i_n}}$$

is again a derivation.[4] *Note that an element $\delta \in \Delta(\mathrm{H}^*)$ is, by definition, zero if it evaluates on H^* to zero: $\delta(h) = 0 \ \forall \ h \in \mathrm{H}^*$.*

The main goal of this section is to generalize in two ways some results of §2 and §5 of [1]:

(1) firstly, to algebras H^* over arbitrary Galois fields,
(2) and secondly by weakening the assumption of J. F. Adams and C. W. Wilkerson that H^* be an integral domain to H^* being reduced, i.e., the nil radical $\mathcal{N}i\ell(\mathrm{H}^*) = (0)$.[5]

We will collect these results in Proposition 1.1.7.

The next convention should make life easier:

CONVENTION: Define the Steenrod operation $\mathcal{P}^i :\equiv 0$ if $i \notin \mathbb{N}_0$.

First we need a generalization of Lemma 2.2 of [1].

LEMMA 1.1.2: *For all natural numbers $i \geq 1$ the following commutation rule holds*

$$\mathcal{P}^k\mathcal{P}^{\Delta_i} - \mathcal{P}^{\Delta_i}\mathcal{P}^k = \mathcal{P}^{\Delta_{i+1}}\mathcal{P}^{k-q^i}.$$

If $i = 0$ we have

$$\mathcal{P}^k\mathcal{P}^{\Delta_0} - \mathcal{P}^{\Delta_0}\mathcal{P}^k = k\mathcal{P}^k.$$

---

[3] Note again the change from the topologist's degree conventions, [25].

[4] Note that the H^*-submodule of the endomorphisms of H^* generated by the Steenrod derivations $\mathcal{P}^{\Delta_i}$, $i \geq 1$ is neither a subalgebra of $\mathcal{P}^*$ nor of the algebra $\mathrm{H}^* \odot \mathcal{P}^*$, see [24] Definition 2.5. In the latter it is just an $\mathbb{F}_q$-vector subspace.

[5] Note that the ground field is always assumed to be an arbitrary Galois field. Assumptions on the existence or non-existence of nilpotent elements or zero divisors or elements of certain degrees in H^* are explicitly stated when needed. In particular, if we need to assume H^* is Noetherian we will explicitly say so.

PROOF: Since $\mathcal{P}^{\Delta_0}$ is just multiplication with the degree of h, the second statement is clear. For the first statement, note that if we had used Milnor's notation for the Steenrod operations, i.e.,

$$\mathcal{P}^k \hat{=} P^{(k,0,\dots)}$$

and

$$\mathcal{P}^{\Delta_i} \hat{=} P^{(0,\dots,0,1,0,\dots)},$$

where the 1 comes in the i-th position, then the proof would have been just an application of Milnor's multiplication rules for Steenrod operations, [25] Theorem 4b. The translation, back and forth, between the two notations is an exercise, that one might look up in the Appendix Lemma A.1.1. •

The second result generalizes Lemma 2.3 of [1] to arbitrary Galois fields. Note the proof is new and completely constructive.

LEMMA 1.1.3: *Let* H^* *be an unstable algebra over* $\mathcal{P}^*$ *and let* $h \in \mathrm{H}^*$ *be an arbitrary element*[6] *of degree* d. *Then for any* $r \geq 0$ *and for any* $s \geq 0$ *there exists an element* $\mathcal{M}_{r,d} \in \mathcal{P}^*$ *such that*

$$\mathcal{P}^{\Delta_{r+s}}(\mathcal{M}_{r,d}(h)) = (\mathcal{P}^{\Delta_s}(h))^{q^r}$$

where the notation $\mathcal{M}_{r,d}$ *emphasizes, that* $\mathcal{M}_{r,d}$ *depends on* r *and* d.

PROOF: Set $d_0 := d = \deg(h)$, and define inductively

$$\begin{aligned}
\mathcal{M}_{0,d} &:= \mathcal{P}^0, \\
\mathcal{M}_{1,d} &:= \mathcal{P}^{d_0-1}, \\
&\vdots \\
\mathcal{M}_{r+1,d} &:= \mathcal{P}^{d_r-1}\mathcal{M}_{r,d}, \qquad \text{where } d_r := \deg(\mathcal{M}_{r,d}(h)).
\end{aligned}$$

Then, it can be proven by induction, with the help of Lemma 1.1.2, that these elements $\mathcal{M}_{r,d}$ satisfy the claimed equation. A detailed proof can be found in the Appendix Lemma A.1.2. •

REMARK: The recursion formulae for the elements $\mathcal{M}_{r,d}$ given in the above proof lead to the following explicit description

$$\mathcal{M}_{r,d} = \mathcal{P}^{d_{r-1}-1} \cdots \mathcal{P}^{d_0-1}\mathcal{P}^0.$$

Moreover, one can prove by induction on r, that (see Appendix Lemma A.1.3)

$$d_r = q^r d_0 - q^r + 1.$$

[6] A careful reading of the proof shows that it would be enough to assume that the element $h \in \mathrm{H}^*$ is unstable,i.e.,

$$\mathcal{P}^i(h) = \begin{cases} h^q & \text{if } i = \deg(h) \\ 0 & \text{if } i > \deg(h). \end{cases}$$

The following series of lemmata lead to the result that, if $m + 1$ derivations $\mathcal{P}^{\Delta_i}$ are linearly dependent, then every $m + 1$ are. Let's start with a recollection of the definition:

DEFINITION: *$D_0, \ldots, D_m \subset \Delta(\mathrm{H}^*)$ are* **linearly dependent**, *if there exist elements $a_0, \ldots, a_m \in \mathrm{H}^*$, not all of which are zero, such that*

$$a_0 D_0 + \cdots + a_m D_m \equiv 0,$$

i.e., evaluating on H^* gives zero; otherwise, call them **linearly independent**.

The following lemma corresponds to Lemma 5.3 of [1]. Note that the assumption that the nil radical be trivial cannot be removed: this is the reason why we will have to distinguish between two cases in the future.

LEMMA 1.1.4: *Let H^* be a reduced unstable algebra, and let $r \in \mathbb{N}_0$. If*

$$\mathcal{P}^{\Delta_{i_0+r}}, \ldots, \mathcal{P}^{\Delta_{i_m+r}}$$

are linearly dependent, then so are

$$\mathcal{P}^{\Delta_{i_0}}, \ldots, \mathcal{P}^{\Delta_{i_m}}.$$

PROOF: Without loss of generality we assume that m is minimal with the property that the derivations $\mathcal{P}^{\Delta_{i_0+r}}, \ldots, \mathcal{P}^{\Delta_{i_m+r}}$ are linearly dependent. By definition, there exist elements $a_0, \ldots, a_m \in \mathrm{H}^*$, not all zero, such that

$$a_0 \mathcal{P}^{\Delta_{i_0+r}} + \cdots + a_m \mathcal{P}^{\Delta_{i_m+r}} \equiv 0$$

on H^*. Hence for any $(m+1)$-tupel $h_0, \ldots, h_m$ of elements of H^* the matrix

$$\left(\mathcal{P}^{\Delta_{i_\alpha+r}}(h_\beta)\right)_{\alpha,\beta=0,\ldots,m}$$

has linearly dependent rows, and therefore its determinant is zero. Lemma 1.1.3 guarantees the existence of elements $\mathcal{M}_{r,d_\beta} \in \mathcal{P}^*$ such that

$$(\mathcal{P}^{\Delta_{i_\alpha}}(h_\beta))^{q^r} = \mathcal{P}^{\Delta_{i_\alpha+r}}(\mathcal{M}_{r,d_\beta}(h_\beta)),$$

where $d_\beta = \deg(h_\beta)$. Therefore

$$\begin{aligned}\left(\det\left(\mathcal{P}^{\Delta_{i_\alpha}}(h_\beta)\right)_{\alpha,\beta=0,\ldots,m}\right)^{q^r} &= \det\left(\mathcal{P}^{\Delta_{i_\alpha}}(h_\beta)^{q^r}\right)_{\alpha,\beta=0,\ldots,m} \\ &= \det\left(\mathcal{P}^{\Delta_{i_\alpha+r}}(\mathcal{M}_{r,d_\beta}(h_\beta))\right)_{\alpha,\beta=0,\ldots,m} \\ &= 0.\end{aligned}$$

Since $\mathcal{N}il(\mathrm{H}^*) = (0)$ this implies

$$(\star) \qquad \det\left(\mathcal{P}^{\Delta_{i_\alpha}}(h_\beta)\right)_{\alpha,\beta=0,\ldots,m} = 0,$$

for every choice of $h_0, \ldots, h_m$. From this we deduce that the derivations $\mathcal{P}^{\Delta_{i_0}}, \ldots, \mathcal{P}^{\Delta_{i_m}}$ are linearly dependent as follows. If $\mathcal{P}^{\Delta_{i_0}}(h_0) = 0$ for all elements $h_0 \in \mathrm{H}^*$, then

$$\mathcal{P}^{\Delta_{i_0}} + 0\mathcal{P}^{\Delta_{i_1}} + \cdots + 0\mathcal{P}^{\Delta_{i_m}} = 0$$

is a relation of linear dependence. Otherwise, choose $k \in \{0, \ldots, m\}$ maximal such that there exist $\overline{h}_0, \ldots, \overline{h}_k \in \mathrm{H}^*$ with

$$\det\left(\mathcal{P}^{\Delta_{i_\alpha}}(\overline{h}_\beta)\right)_{\alpha,\beta=0,\ldots,k} \neq 0.$$

By (★) we have that $k < m$, and for every $h_0, \ldots, h_m \in \mathrm{H}^*$, we have[7]

$$\begin{aligned} 0 &= \det\left(\mathcal{P}^{\Delta_{i_\alpha}}(h_\beta)\right)_{\alpha,\beta=0,\ldots,k+1} + 0\mathcal{P}^{\Delta_{i_{k+2}}} + \cdots + 0\mathcal{P}^{\Delta_{i_m}} \\ &= \sum_{j=0}^{k+1} (-1)^{k+2+j} \det\left(\mathcal{P}^{\Delta_{i_\alpha}}(h_\beta)\right)^{\beta=0,\ldots,k}_{\alpha,=0,\ldots,\widehat{j},\ldots,k+1} \mathcal{P}^{\Delta_{i_j}}(h_{k+1}) \quad \forall\, h_{k+1} \in \mathrm{H}^*. \end{aligned}$$

The last equation follows from Lagrange expansion with respect to the last column. Since by choice of k the coefficient of $\mathcal{P}^{\Delta_{i_{k+1}}}$ is not zero, this equation gives a nontrivial relation of linear dependence between $\mathcal{P}^{\Delta_{i_0}}, \ldots, \mathcal{P}^{\Delta_{i_m}}$. •

Our next lemma generalizes Lemma 5.4 of [1].

LEMMA 1.1.5: *Let H^* be a reduced unstable algebra, and let $r \in \mathbb{N}_0$. If*

$$\mathcal{P}^{\Delta_0}, \mathcal{P}^{\Delta_{i_1}}, \ldots, \mathcal{P}^{\Delta_{i_m}}$$

are linearly dependent, where $i_j > 0$ for $j = 1, \ldots, m$, then, for every $r \geq 0$, so are

$$\mathcal{P}^{\Delta_r}, \mathcal{P}^{\Delta_{i_1}}, \ldots, \mathcal{P}^{\Delta_{i_m}}.$$

PROOF: The statement is redundant if $r = 0$. So assume that $r > 0$. Let $h_0, \ldots, h_m \in \mathrm{H}^*$ such that

$$h_0\mathcal{P}^{\Delta_0} + h_1\mathcal{P}^{\Delta_{i_1}} + \cdots + h_m\mathcal{P}^{\Delta_{i_m}} = 0$$

is a non-trivial linear relation. Assume without loss of generality that h_0 is a q-th power (if not multiply the above equation by h_0^{q-1}). From the commutation rules for the derivations $\mathcal{P}^{\Delta}$'s, see page 12, we get

$$\begin{aligned} 0 &= \mathcal{P}^{\Delta_r}\left(h_0\mathcal{P}^{\Delta_0} + h_1\mathcal{P}^{\Delta_{i_1}} + \cdots + h_m\mathcal{P}^{\Delta_{i_m}}\right) \\ &= \mathcal{P}^{\Delta_r}(h_0)\mathcal{P}^{\Delta_0} + \mathcal{P}^{\Delta_r}(h_1)\mathcal{P}^{\Delta_{i_1}} + \cdots + \mathcal{P}^{\Delta_r}(h_m)\mathcal{P}^{\Delta_{i_m}} \\ &\qquad + h_0\mathcal{P}^{\Delta_0}\mathcal{P}^{\Delta_r} + h_1\mathcal{P}^{\Delta_{i_1}}\mathcal{P}^{\Delta_r} + \cdots + h_m\mathcal{P}^{\Delta_{i_m}}\mathcal{P}^{\Delta_r} + h_0\mathcal{P}^{\Delta_r} \\ &= h_0\mathcal{P}^{\Delta_r} + \mathcal{P}^{\Delta_r}(h_1)\mathcal{P}^{\Delta_{i_1}} + \cdots + \mathcal{P}^{\Delta_r}(h_m)\mathcal{P}^{\Delta_{i_m}}, \end{aligned}$$

[7] We adopt the topologist's convention that a ^ over an element means that it is omitted.

where the last equality holds, because

$$h_0\mathcal{P}^{\Delta_0}\mathcal{P}^{\Delta_r} + h_1\mathcal{P}^{\Delta_{i_1}}\mathcal{P}^{\Delta_r} + \cdots + h_m\mathcal{P}^{\Delta_{i_m}}\mathcal{P}^{\Delta_r} = \left(h_0\mathcal{P}^{\Delta_0} + h_1\mathcal{P}^{\Delta_{i_1}} + \cdots + h_m\mathcal{P}^{\Delta_{i_m}}\right)\mathcal{P}^{\Delta_r} \equiv 0$$

by assumption, and $\mathcal{P}^{\Delta_r}(h_0) = 0$, since h_0 is a q-th power. Therefore

$$\mathcal{P}^{\Delta_r},\ \mathcal{P}^{\Delta_{i_1}}, \ldots,\ \mathcal{P}^{\Delta_{i_m}}$$

are linearly dependent. •

PROPOSITION 1.1.6: *Let* H^* *be a reduced unstable algebra. Let the derivations* $\mathcal{P}^{\Delta_{i_0}}, \ldots, \mathcal{P}^{\Delta_{i_m}}$ *be linearly dependent, then so are*

$$\mathcal{P}^{\Delta_0}, \ldots, \mathcal{P}^{\Delta_m}.$$

PROOF: Denote by $\mathcal{S}$ the set of all $(m + 1)$-tuples $(s_0, \ldots, s_m)$ such that

$$\mathcal{P}^{\Delta_{s_0}}, \ldots, \mathcal{P}^{\Delta_{s_m}}$$

are linearly dependent. By assumption this set in not empty. We have to show that

$$(0, \ldots, m) \in \mathcal{S}.$$

We will do this by giving an algorithm for the construction of a relation of linear dependence for $\mathcal{P}^{\Delta_0}, \ldots, \mathcal{P}^{\Delta_m}$ starting from an arbitrary $(i_0, \ldots, i_m) \in \mathcal{S}$.

By Lemma 1.1.4 we know that

$$(0,\ i_1 - i_0, \ldots,\ i_m - i_0) \in \mathcal{S}.$$

STEP 1: If $(0,\ i_1 - i_0, \ldots,\ i_m - i_0) \neq (0, \ldots, m)$ define

$$\alpha := \min\{1, \ldots, m \mid i_\alpha - i_{\alpha-1} > 1\},$$

i.e.,

$$(0,\ 1, \ldots,\ \alpha - 1,\ i_\alpha - i_0, \ldots,\ i_m - i_0) \in \mathcal{S}.$$

STEP 2: Choosing $r = \alpha$, Lemma 1.1.5 gives, after reordering, that

$$(1, \ldots,\ \alpha,\ i_\alpha - i_0, \ldots,\ i_m - i_0) \in \mathcal{S}.$$

STEP 3: Applying Lemma 1.1.4 with $r = 1$ leads to

$$(0,\ldots, \alpha-1,\ i_\alpha - i_0 - 1,\ldots, i_m - i_0 - 1) \in \mathcal{S}.$$

If $(0,\ldots, \alpha - 1,\ i_\alpha - i_0 - 1,\ldots, i_m - i_0 - 1) = (0,\ldots, m)$ we are done. If not, either

$$(i_\alpha - i_0 - 1) - (\alpha - 1) > 1, \quad \text{or} \quad (i_\alpha - i_0 - 1) - (\alpha - 1) = 1.$$

In the first case repeat the Steps 2 and 3, in the latter, the "gap" between the ascending numbers i_j has moved to the right. So, define α renewed, as explained in Step 1, and repeat the procedure. In either, case we eventually get,

$$(0,\ldots, m) \in \mathcal{S}$$

as claimed. •

We summarize our results:

PROPOSITION 1.1.7: *Let H^* be an unstable algebra over the Steenrod algebra. Let $\mathcal{N}il(\mathrm{H}^*) = (0)$. If there is an $(m+1)$-tuple of linearly dependent derivations $\mathcal{P}^{\Delta_{i_0}},\ldots, \mathcal{P}^{\Delta_{i_m}}$, then any $m+1$ are linearly dependent.*

PROOF: Let $(i_0,\ldots, i_m) \in \mathcal{S}$. Then by Proposition 1.1.6 $(0,\ldots, m) \in \mathcal{S}$. From this we show that any $(m+1)$-tuple $(k_0,\ldots, k_m)$ belongs to $\mathcal{S}$. Define

$$r_i := k_i + (m - i).$$

By assumption $(0,\ldots, m) \in \mathcal{S}$.

STEP 1: Define $r = r_0$ and apply Lemma 1.1.5. This gives

$$(r_0,\ 1,\ldots, m) \in \mathcal{S}.$$

STEP 2: Define $r = 1$ and apply Lemma 1.1.4. Then we get

$$(r_0 - 1,\ 0,\ldots, m-1) \in \mathcal{S}.$$

Repeat Step 1 with $r := r_1$, and then Step 2. We get

$$(r_0 - 2,\ r_1 - 1,\ 0,\ldots, m-2) \in \mathcal{S}.$$

After doing this m times we find

$$(r_0 - m,\ r_1 - (m-1),\ldots, r_{m-1} - 1,\ 0) \in \mathcal{S}.$$

Apply Lemma 1.1.5 once more with $r = r_m$ to give

$$(k_0,\ldots, k_m) = (r_0 - m,\ r_1 - (m-1),\ldots, r_m) \in \mathcal{S}.$$

This is the desired algorithm. •

The following result generalizes Lemma 5.5 of [1]. We will need this later in Chapter 5.

LEMMA 1.1.8: *Let* H^* *be a reduced unstable algebra. If, for any elements* $a_1, \ldots, a_m \in \mathrm{H}^*$,

$$0 \equiv \sum_{i=0}^{m} h_i \mathcal{P}^{\Delta_i} = \det \begin{bmatrix} \mathcal{P}^{\Delta_0} & \mathcal{P}^{\Delta_0}(a_1) & \cdots & \mathcal{P}^{\Delta_0}(a_m) \\ \vdots & \vdots & \ddots & \vdots \\ \mathcal{P}^{\Delta_m} & \mathcal{P}^{\Delta_m}(a_1) & \cdots & \mathcal{P}^{\Delta_m}(a_m) \end{bmatrix}$$

evaluates to zero on H^*, *then*

$$h_0^{q^r} \mathcal{P}^{\Delta_r} + \cdots + h_m^{q^r} \mathcal{P}^{\Delta_{r+m}} \equiv 0$$

is zero in $\Delta(\mathrm{H}^*)$ *for any* $r \in \mathbb{N}_0$.

PROOF: From what we have proved so far we know that for any $r \in \mathbb{N}_0$

$$\mathcal{P}^{\Delta_r}, \ldots, \mathcal{P}^{\Delta_{r+m}}$$

are linearly dependent as elements in $\Delta(\mathrm{H}^*)$, provided $\mathcal{P}^{\Delta_0}, \ldots, \mathcal{P}^{\Delta_m}$ are. Without loss of generality let m be minimal with this property. In particular, there exist elements $k_0, \ldots, k_m \in \mathrm{H}^*$ such that

$$k_0 \mathcal{P}^{\Delta_r} + \cdots + k_m \mathcal{P}^{\Delta_{r+m}} = 0 \in \Delta(\mathrm{H}^*).$$

We have to determine the k_i's.

First, note that Lemma 1.1.3 guarantees the existence of an element $\mathcal{M}_{r,\,d} \in \mathcal{P}^*$ depending on r and $d = \deg(h)$, such that

$$\mathcal{P}^{\Delta_{r+i}}\left(\mathcal{M}_{r,\,d}(h)\right) = \left(\mathcal{P}^{\Delta_i}(h)\right)^{q^r}.$$

Choose $a_1, \ldots, a_m \in \mathrm{H}^*$ such that

$$\det\left(\mathcal{P}^{\Delta_i}(a_j)\right)_{i,\,j=1,\ldots,m} \neq 0.$$

Such a choice is possible since m is minimal by assumption. Therefore, for $d_j = \deg(a_j)$,

$$\det\left(\mathcal{P}^{\Delta_{r+i}}\left(\mathcal{M}_{r,\,d_j}(a_j)\right)\right)_{i,\,j=1,\ldots,m} = \left(\det\left(\mathcal{P}^{\Delta_i}(a_j)\right)_{i,\,j=1,\ldots,m}\right)^{q^r} \neq 0.$$

Consider the $(m+1) \times m$ matrix

$$\mathbf{M} := \left(\mathcal{P}^{\Delta_{r+i}}(\mathcal{M}_{r,\,d_j}(a_j))\right)_{i=0,\ldots,m,\ j=1,\ldots,m}$$
$$= \begin{bmatrix} \mathcal{P}^{\Delta_r}(\mathcal{M}_{r,\,d_1}(a_1)) & \cdots & \mathcal{P}^{\Delta_r}(\mathcal{M}_{r,\,d_m}(a_m)) \\ \vdots & \ddots & \vdots \\ \mathcal{P}^{\Delta_{r+m}}(\mathcal{M}_{r,\,d_1}(a_1)) & \cdots & \mathcal{P}^{\Delta_{r+m}}(\mathcal{M}_{r,\,d_m}(a_m)) \end{bmatrix}$$

and denote by $\mathbf{M}_i$ the $m \times m$ submatrix which is obtained from $\mathbf{M}$ by erasing the $(i+1)$-st row. From the above we have

$$\det(\mathbf{M}_i) \neq 0$$

for $i = 0, \dots, m$. Setting $k_i := \det(\mathbf{M}_i)$ we have, by Lagrange expansion using the first column,

$$\sum_{i=0}^{m} k_i \mathcal{P}^{\Delta_{r+i}} = \det \begin{bmatrix} \mathcal{P}^{\Delta_r} & \mathcal{P}^{\Delta_r}(\mathcal{M}_{r,\, d_1}(a_1)) & \cdots & \mathcal{P}^{\Delta_r}(\mathcal{M}_{r,\, d_m}(a_m)) \\ \vdots & \vdots & \ddots & \vdots \\ \mathcal{P}^{\Delta_{r+m}} & \mathcal{P}^{\Delta_{r+m}}(\mathcal{M}_{r,\, d_1}(a_1)) & \cdots & \mathcal{P}^{\Delta_{r+m}}(\mathcal{M}_{r,\, d_m}(a_m)) \end{bmatrix} = 0,$$

where the last equality follows because $\mathcal{P}^{\Delta_r}, \dots, \mathcal{P}^{\Delta_{r+m}}$ are linearly dependent. Hence

$$\sum_{i=0}^{m} k_i \mathcal{P}^{\Delta_{r+i}} = \det \begin{bmatrix} \mathcal{P}^{\Delta_r} & (\mathcal{P}^{\Delta_0}(a_1))^{q^r} & \cdots & (\mathcal{P}^{\Delta_0}(a_m))^{q^r} \\ \vdots & \vdots & \ddots & \vdots \\ \mathcal{P}^{\Delta_{r+m}} & (\mathcal{P}^{\Delta_m}(a_1))^{q^r} & \cdots & (\mathcal{P}^{\Delta_m}(a_m))^{q^r} \end{bmatrix}$$

$$= \sum_{i=0}^{m} h_i^{q^r} \mathcal{P}^{\Delta_{r+i}},$$

since taking to q-th powers is additive over $\mathbb{F}_q$. •

1.2 The Δ-Theorem, Corollaries and Examples

We are now ready to prove the promised highlight of this chapter: the Δ-Theorem. In addition, we will draw some consequences and exhibit some examples.

First recall Proposition 1.1.7: if $m+1$ $\mathcal{P}^{\Delta}$'s of the H^*-module of derivations $\Delta(\mathrm{H}^*)$ are linearly dependent and, if $\mathcal{N}il(\mathrm{H}^*) = (0)$, then the $m+1$ derivations of lowest degree

$$\mathcal{P}^{\Delta_0}, \dots, \mathcal{P}^{\Delta_m}$$

are linearly dependent.[8] This motivates the following definition.

DEFINITION: *Let* H^* *be an unstable algebra over the Steenrod algebra with*[9] $\mathcal{N}il(\mathrm{H}^*) = (0)$, *and let* $\Delta(\mathrm{H}^*)$ *be the* H^**-module of derivations generated by* $\mathcal{P}^{\Delta_i}$, $\forall\, i \in \mathbb{N}_0$. *If there exists an* $m \in \mathbb{N}_0$ *such that some* $m+1$ *derivations (and hence by Proposition* 1.1.7 *any* $m+1$ *derivations) are linearly dependent, then* H^* *is called* Δ**-finite**. *A relation*[10] *of linear dependence*

$$h_0 \mathcal{P}^{\Delta_0} + \cdots + h_m \mathcal{P}^{\Delta_m} = 0$$

[8] N.B.: This does *not* imply that $\Delta(\mathrm{H}^*)$ is finitely generated as an H^*-module.

[9] We make the assumption that H^* is reduced, because we want to use Proposition 1.1.7. However, the definition makes sense without this assumption.

[10] Note that the coefficients $h_0, \dots, h_m$ are in general not unique, but the Δ-length $m(\mathrm{H}^*)$ *is*, by minimality.

(where m is chosen to be minimal w.r.t. this property) is called a **Δ-relation** *of* H^* *and* $m = m(\mathrm{H}^*)$ *the* **Δ-length**.

Let's have a first example.

EXAMPLE 1: Let $\mathrm{H}^* = \mathbb{F}[x]$ be a polynomial algebra in one linear generator over $\mathbb{F} = \mathbb{F}_q$. Then H^* is Δ-finite, because

$$-x^{q-1}\mathcal{P}^{\Delta_0} + \mathcal{P}^{\Delta_1} = 0.$$

Moreover, as Lemma 1.1.8 predicts, we have

$$-x^{(q-1)q^i}\mathcal{P}^{\Delta_i} + \mathcal{P}^{\Delta_{i+1}} = 0.$$

This leads inductively to

$$\mathcal{P}^{\Delta_i} \in \mathbb{F}[x]\mathcal{P}^{\Delta_0} \quad \forall\, i \in \mathbb{N}_0,$$

i.e.,

$$\Delta(\mathbb{F}[x]) = \mathbb{F}[x]\mathcal{P}^{\Delta_0} \subset \mathrm{End}(\mathbb{F}[x])$$

is a (even free!) $\mathbb{F}[x]$-module generated by the single derivation of lowest degree $\mathcal{P}^{\Delta_0}$. Much, much, later we will see that this is not an isolated example, see Proposition 8.2.2.

The following theorem shows that for Noetherian algebras the module of derivations is always finitely generated, so in particular, if we add the assumption that our algebra is reduced, it is also Δ-finite. This generalizes Theorem 5.1 of [1] firstly, to algebras over arbitrary Galois fields and, secondly, to algebras, which can have zero divisors (see also [36] Theorem 10.5.4). For the latter we have to make a fine distinction between the statement that $\Delta(\mathrm{H}^*)$ is finitely generated and Δ-finiteness of H^*.

THEOREM 1.2.1 (Δ-Theorem): *Let* H^* *be a Noetherian unstable algebra over the Steenrod algebra* $\mathcal{P}^*$. *Then* $\Delta(\mathrm{H}^*)$ *is finitely generated as an* H^*-*module. If in addition the nil radical of* H^* *is trivial, i.e.,* $\mathcal{N}il(\mathrm{H}^*) = (0)$, *then* H^* *is* Δ-*finite, i.e., there exist a natural number* $m \in \mathbb{N}_0$ *and non-zero elements* $h_0, \ldots, h_m \in \mathrm{H}^*$ *such that*

$$h_0\mathcal{P}^{\Delta_0} + \cdots + h_m\mathcal{P}^{\Delta_m} = 0,$$

where m *is minimal with respect to this property.*

PROOF: Let $y_1, \ldots, y_s$ be generators of H^* as an algebra over $\mathbb{F}_q$. Then we have a monomorphism of H^*-modules defined by

$$\begin{array}{rccc} \varphi: & \Delta(\mathrm{H}^*) & \hookrightarrow & \bigoplus_s \mathrm{H}^* \\ & \delta & \longmapsto & (\delta(y_1), \ldots, \delta(y_s)). \end{array}$$

Injectivity follows from:

$$\varphi(\delta) = 0 \iff \delta(\mathrm{H}^*) \equiv 0 \iff \delta = 0,$$

by definition of $\Delta(\mathrm{H}^*)$. Hence $\Delta(\mathrm{H}^*)$ is isomorphic to its image $\varphi(\Delta(\mathrm{H}^*))$, which in turn is a submodule of a finitely generated Noetherian H^*-module, namely $\bigoplus_s \mathrm{H}^*$. Therefore $\Delta(\mathrm{H}^*)$ is also finitely generated as an H^*-module, i.e., there exists a natural number, $g = g(\mathrm{H}^*)$, such that $\Delta(\mathrm{H}^*)$ is generated by

$$\delta_1, \ldots, \delta_g.$$

Without loss of generality we can assume that the δ's are $\mathcal{P}^{\Delta}$'s, so we get a generating set consisting of

$$\mathcal{P}^{\Delta_{i_1}}, \ldots, \mathcal{P}^{\Delta_{i_g}}.$$

This, in turn, means that any other element $\mathcal{P}^{\Delta_i}$ in our module can be expressed as a linear combination of these, in particular, for $i \in \mathbb{N}_0$

$$\mathcal{P}^{\Delta_i} = a_1 \mathcal{P}^{\Delta_{i_1}} + \cdots + a_g \mathcal{P}^{\Delta_{i_g}}$$

for some $a_1, \ldots, a_g \in \mathrm{H}^*$. In other words: for any $i \notin \{i_1, \ldots, i_g\}$ the elements

$$\mathcal{P}^{\Delta_i}, \mathcal{P}^{\Delta_{i_1}}, \ldots, \mathcal{P}^{\Delta_{i_g}}$$

are linearly dependent. Let m be minimal with respect to the property that $\mathcal{P}^{\Delta_{i_0}}, \ldots, \mathcal{P}^{\Delta_{i_m}}$ are linearly dependent, for some $i_0, \ldots, i_m \in \mathbb{N}_0$. Then, by Proposition 1.1.6, we get that[11]

$$\mathcal{P}^{\Delta_0}, \ldots, \mathcal{P}^{\Delta_m}$$

are linearly dependent. Hence H^* is Δ-finite and we are all set. •

However, Noetherianess is not necessary for being Δ-finite. The next example illustrates this.

EXAMPLE 2: Let $\mathbb{F}_2$ be the field with two elements. Take the polynomial algebra in two linear generators, $\mathbb{F}_2[x, y]$, and let H^* be the subalgebra generated by $x, xy, xy^2, xy^3, \ldots$. Then H^* is not Noetherian, but certainly reduced. However, H^* is Δ-finite with Δ-relation

$$x(xy^2 + x^2y)\mathcal{P}^{\Delta_0} + x(x^2 + xy + y^2)\mathcal{P}^{\Delta_1} + x\mathcal{P}^{\Delta_2} = 0$$

of length 2. Note carefully, that the coefficients are multiples of the Dickson classes of $\mathbb{F}_2[x, y]$

$$\begin{aligned} x\mathbf{d}_{2,0} &= x(xy^2 + x^2y) \\ x\mathbf{d}_{2,1} &= x(x^2 + xy + y^2) \\ x\mathbf{d}_{2,2} &= x, \end{aligned}$$

where $\mathbf{d}_{n,n} = 1, \ \forall\, n$, by convention.

---

[11] It is exactly here, that the assumption that the nil radical of H^* is zero, comes into the game: Proposition 1.1.6 depends on Lemma 1.1.3, which in turn needs precisely this: the non-existence of nilpotent elements.

REMARK: Since, for $i \geq 0$,

$$\deg(\mathcal{P}^{\Delta_i}) = q^i - 1,$$

the homogeneity of a Δ-relation leads to

$$\deg(h_i) = q^m - q^i + \deg(h_m),$$

for any $i = 0, \dots, m-1$ as elements of the graded algebra H^*.

REMARK: If $h_m = 1$, or more generally if h_m is a unit, then the coefficients $h_0, \dots, h_{m-1}$ are uniquely determined by minimality, so, in this case it is justified to use the expression *the* Δ-relation.

Let's have a look at another example.

EXAMPLE 3: Take the field $\mathbb{F}_2$ of two elements and consider a polynomial algebra in two linear generators x, y over $\mathbb{F}_2$, $\mathbb{F}[x, y]$. We have a chain of unstable algebras

$$\mathbb{F}_2[x^2, y^2] \hookrightarrow \mathbb{F}_2[x^2, y] \hookrightarrow \mathbb{F}_2[x, y].$$

The Δ-relation in the smallest of these, $\mathbb{F}_2[x^2, y^2]$, has length 0

$$\mathcal{P}^{\Delta_0} = 0.$$

For the middle algebra, $\mathbb{F}_2[x^2, y]$, we get a Δ-relation of length 1

$$y\mathcal{P}^{\Delta_0} + \mathcal{P}^{\Delta_1} = 0.$$

For the big algebra, $\mathbb{F}_2[x, y]$, our Δ-relation has length 2

$$(x^2y + xy^2)\mathcal{P}^{\Delta_0} + (x^2 + xy + y^2)\mathcal{P}^{\Delta_1} + \mathcal{P}^{\Delta_2} = 0.$$

Note that the length of the respective Δ-relations does not exceed the Krull dimension. This is true in a more general setting:

COROLLARY 1.2.2: *Let* H^* *be a reduced Noetherian*[12] *unstable algebra over the Steenrod algebra. Then the Krull dimension of* H^*, $\dim(H^*)$, *is at least* $m = m(H^*)$, *the* Δ*-length.*

PROOF: Let $\dim(H^*) = n$. Then any set of $n + 1$ elements $a_0, \dots, a_n \in H^*$ is algebraically dependent. Hence the determinant of the generalized Jacobian matrix

$$\det\left(\mathcal{P}^{\Delta_i}(a_j)\right)_{i,\, j=0,\dots,n} = \bar{h}_0\mathcal{P}^{\Delta_0}(a_0) + \cdots + \bar{h}_n\mathcal{P}^{\Delta_n}(a_0)$$

vanishes or is a zero divisor for every $a_0 \in H^*$, see Theorem A.4.1 in the Appendix. The last equation follows using the Lagrange expansion of the first column. Hence for some element $\bar{h} \in H^* \setminus \{0\}$ we get a relation of the form

$$0 = \bar{h}\det\left(\mathcal{P}^{\Delta_i}(a_j)\right)_{i,\, j=0,\dots,n} = \bar{h}\left(\bar{h}_0\mathcal{P}^{\Delta_0}(a_0) + \cdots + \bar{h}_n\mathcal{P}^{\Delta_n}(a_0)\right)$$

[12] What remains true if we replace Noetherianess by the weaker assumption of Δ-finiteness is investigated in [27].

However, if this is the trivial relation, i.e.,

$$\bar{h}\bar{h}_0 = \cdots = \bar{h}\bar{h}_n = 0,$$

then, since

$$\bar{h}_k = \det\left(\mathcal{P}^{\Delta_i}(a_j)\right)_{i,\ j=0,\dots,\widehat{k},\dots,n},$$

where we adopt the topologist's convention that $\widehat{\ }$ means that this element is omitted, the generalized Jacobian matrix

$$\bar{h}_n = \det\left(\mathcal{P}^{\Delta_i}(a_j)\right)_{i,\ j=0,\dots,n-1}$$

vanishes or is a zero divisor. So, we would have by Lagrange expansion a relation of length at most $n-1$

$$\tilde{h}\left(\tilde{h}_0\mathcal{P}^{\Delta_0}(a_0) + \cdots + \tilde{h}_{n-1}\mathcal{P}^{\Delta_{n-1}}(a_0)\right) = 0,$$

for all $a_0 \in \mathrm{H}^*$. Proceeding this way we end up with a non-trivial[13] Δ-relation

$$\begin{aligned} 0 &= \check{h}\det\left(\mathcal{P}^{\Delta_i}(a_j)\right)_{i,\ j=0,\dots,k} \\ &= \check{h}\left(\check{h}_0\mathcal{P}^{\Delta_0}(a_0) + \cdots + \check{h}_k\mathcal{P}^{\Delta_k}(a_0)\right) \end{aligned}$$

for some k, $0 \le k \le n$. •

Recall the example above, $\mathbb{F}_2[x,\ y]$, and its Δ-relation

$$(x^2y + xy^2)\mathcal{P}^{\Delta_0} + (x^2 + xy + y^2)\mathcal{P}^{\Delta_1} + \mathcal{P}^{\Delta_2} \equiv 0.$$

Note that the coefficients are exactly the Dickson classes

$$\begin{aligned} \mathbf{d}_{2,0} &= x^2y + xy^2 \\ \mathbf{d}_{2,1} &= x^2 + xy + y^2. \end{aligned}$$

This is no accident as the next theorem shows, compare [36] Proposition 10.4.3 for the case of prime fields.

THEOREM 1.2.3: *The Δ-relation for $\mathbb{F}[x_1, \dots, x_n]$ is given by*

$$(-1)^n\mathbf{d}_{n,0}\mathcal{P}^{\Delta_0} + \cdots + (-1)\mathbf{d}_{n,n-1}\mathcal{P}^{\Delta_{n-1}} + \mathcal{P}^{\Delta_n} \equiv 0.$$

---

[13] Note that, in worst case, we end up with

$$\check{h}\mathcal{P}^{\Delta_0}(a) = 0 \quad \forall\ a \in \mathrm{H}^*$$

for a suitable $\check{h} \in \mathrm{H}^* \setminus 0$.

PROOF: It is enough to show that

$$(-1)^n \mathbf{d}_{n,0}\mathcal{P}^{\Delta_0}(x_i) + \cdots + (-1)\mathbf{d}_{n,n-1}\mathcal{P}^{\Delta_{n-1}}(x_i) + \mathcal{P}^{\Delta_n}(x_i) = 0$$

for all algebra generators $x_i \in \mathbb{F}[x_1, \ldots, x_n]$. Since our generators have degree 1 we have

$$\mathcal{P}^{\Delta_j}(x_i) = x_i^{q^j}, \quad \forall\, i = 1, \ldots, n, \ \forall\, j \geq 0.$$

Hence

$$\det\left(\mathcal{P}^{\Delta_j}(x_i)\right)_{i,\,j} = \det\left(x_i^{q^j}\right)_{i,\,j}.$$

In [9] L. E. Dickson showed that

$$\det\left(x_i^{q^j}\right)_{i=1,\ldots,n,\ j=0,\ldots,\widehat{k},\ldots,n} = \mathbf{E}_n \mathbf{d}_{n,k},$$

where $\mathbf{E}_n$ is a product[14] of linear forms, one for each one-dimensional vector subspace in V^*. Therefore we get for every $l = 1, \ldots, n$

$$\begin{aligned}
&\mathbf{E}_n \left[(-1)^n \mathbf{d}_{n,0}\left(\mathcal{P}^{\Delta_0}(x_l)\right) + \cdots + (-1)\mathbf{d}_{n,n-1}\left(\mathcal{P}^{\Delta_{n-1}}(x_l)\right) + \left(\mathcal{P}^{\Delta_n}(x_l)\right)\right] \\
&= (-1)^n \Big[\mathbf{E}_n \mathbf{d}_{n,0}\mathcal{P}^{\Delta_0}(x_l) + \cdots \\
&\qquad \cdots + (-1)^{n-1}\mathbf{E}_n \mathbf{d}_{n,n-1}\mathcal{P}^{\Delta_{n-1}}(x_l) + (-1)^n \mathbf{E}_n \mathcal{P}^{\Delta_n}(x_l)\Big] \\
&= (-1)^n \left[\det\left(x_i^{q^j}\right)_{i=1,\ldots,n}^{j=1,\ldots,n}\left(\mathcal{P}^{\Delta_0}(x_l)\right) + \cdots\right. \\
&\qquad \cdots + (-1)^{n-1}\det\left(x_i^{q^j}\right)_{i=1,\ldots,n}^{j=0,\ldots,n-2,\,n}\left(\mathcal{P}^{\Delta_{n-1}}(x_l)\right) \\
&\qquad \left. + (-1)^n \det\left(x_i^{q^j}\right)_{i=1,\ldots,n}^{j=0,\ldots,n-1}\left(\mathcal{P}^{\Delta_n}(x_l)\right)\right] \\
&= (-1)^n \left[\det\left(x_i^{q^j}\right)_{i=1,\ldots,n}^{j=1,\ldots,n} x_l + \cdots + (-1)^{n-1}\det\left(x_i^{q^j}\right)_{i=1,\ldots,n}^{j=0,\ldots,n-2,\,n} x_l^{q^{n-1}}\right. \\
&\qquad \left. + (-1)^n \det\left(x_i^{q^j}\right)_{i=1,\ldots,n}^{j=0,\ldots,n-1} x_l^{q^n}\right] \\
&= (-1)^n \det\begin{bmatrix} x_l & x_1 & \cdots & x_n \\ x_l^q & x_1^q & \cdots & x_n^q \\ \vdots & \vdots & \ddots & \vdots \\ x_l^{q^n} & x_1^{q^n} & \cdots & x_n^{q^n} \end{bmatrix}
\end{aligned}$$

[14] Using the convention $\mathbf{d}_{n,n} = 1$, Dickson's formula gives

$$\mathbf{E}_n = \det\left(x_i^{q^j}\right)_{i=1,\ldots,n,\ j=0,\ldots,n-1}.$$

This polynomial is, by the way, the Euler class of the $\mathrm{SL}(n, \mathbb{F})$-action on $V^* \setminus 0$, see Section 8.1 in [36] or [39].

$= 0,$

where we made use of Lagrange expansion with respect to the first column and the assumption that $l \in \{1, \dots, n\}$. Since $\mathbf{E}_n \neq 0$ and $\mathbb{F}[x_1, \dots, x_n]$ is an integral domain, we are done. •

REMARK: Note that the Δ-length of the Dickson algebra, $\mathcal{D}^*(n)$, is n, because otherwise there would exist m elements $a_0, \dots, a_m \in \mathcal{D}^*(n)$, with $m \leq n-1$ and not all zero, such that

$$a_0 \mathcal{P}^{\Delta_0}(\mathbf{d}_{n,i}) + \cdots + a_m \mathcal{P}^{\Delta_m}(\mathbf{d}_{n,i}) = 0$$

for any $i = 0, \dots, n-1$. However, we know how the Steenrod algebra acts on the Dickson polynomials (compare Proposition A.2.1 in the Appendix); in particular

$$\begin{aligned} 0 &= a_0 \mathcal{P}^{\Delta_0}(\mathbf{d}_{n,i}) + \cdots + a_m \mathcal{P}^{\Delta_m}(\mathbf{d}_{n,i}) \\ &= -1 a_i \mathbf{d}_{n,0} \quad \forall\, i = 0, \dots, m. \end{aligned}$$

This means that all the coefficients $a_0, \dots, a_m$ must be zero, which contradicts our assumption. Hence the Δ-relation given in the above theorem is also the Δ-relation[15] for the Dickson algebra $\mathcal{D}^*(n)$, and hence for any unstable algebra between $\mathcal{D}^*(n)$ and $\mathbb{F}[V]$. Note that in Example 1 the Dickson algebra $\mathcal{D}^*(2)$ is not contained in $\mathbb{F}[x^2, y^2]$ nor in $\mathbb{F}[x^2, y]$.

We will need the following proposition later.[16]

PROPOSITION 1.2.4: *Let* H^* *be a* Δ*-finite, reduced, unstable algebra. Let* $h_0, \dots, h_m$ *be the coefficients occuring in a* Δ*-relation. Then, for any* $\alpha \in \mathbb{N}_0$,

$$\Delta(\alpha, h_0, \dots, h_m) := \mathcal{P}^{\alpha}(h_0)\mathcal{P}^{\Delta_0} + \cdots + \mathcal{P}^{\alpha}(h_m)\mathcal{P}^{\Delta_m} + \sum_{i=0}^{m} \mathcal{P}^{\alpha - q^i}(h_i)\mathcal{P}^{\Delta_{i+1}} = 0$$

as an element of the H^**-module* $\Delta(\mathrm{H}^*)$.

PROOF: The statement for $\alpha = 0$ is the contents of the Δ-Theorem 1.2.1, i.e., we have $\Delta(0, h_0, \dots, h_m) = 0$ in $\Delta(\mathrm{H}^*)$, because[17] $\mathcal{P}^0$ is the identity. We proceed by induction on α. So, let $\alpha = 1$, then[18]

$$0 = \mathcal{P}^1 \Delta_0$$

---

15 Since the leading coefficient is a unit, the Δ-relation is indeed unique, compare with the remark after the Δ-Theorem 1.2.1.

16 I agree these calculations at the high school algebra level are not really exciting; but be a bit more patient: we are almost there.

17 We will omit the $h_0, \dots, h_m$ in parentheses and write the α as a subscript, $\Delta_\alpha := \Delta(\alpha, h_0, \dots, h_m)$, in the future whenever there is no chance for confusion.

18 I know that with $\alpha = 0$ our induction started already. However, the case $\alpha = 1$ should convince you that you gonna need a huge piece of paper, a dozen sharp pencils and lots of stamina to do the general case.

$$
\begin{aligned}
&= \sum_{i=0}^{m} \mathcal{P}^1(h_i)\mathcal{P}^{\Delta_i} + h_i\mathcal{P}^1\mathcal{P}^{\Delta_i} \\
&= \sum_{i=0}^{m} \left(\mathcal{P}^1(h_i)\mathcal{P}^{\Delta_i} + h_i\mathcal{P}^{\Delta_i}\mathcal{P}^1\right) + h_0\mathcal{P}^1 \\
&= \left(\sum_{i=0}^{m} \mathcal{P}^1(h_i)\mathcal{P}^{\Delta_i} + h_0\mathcal{P}^1\right) + \sum_{i=0}^{m} h_i\mathcal{P}^{\Delta_i}\mathcal{P}^1 \\
&= \Delta_1 + \Delta_0\mathcal{P}^1,
\end{aligned}
$$

with a little help from the Cartan formulae and Lemma 1.1.2. Since $\Delta_0\mathcal{P}^1 = 0$, the statement is true for $\alpha = 1$.

Let $\alpha > 1$. Then

$$
\begin{aligned}
0 &= \mathcal{P}^\alpha \Delta_0 \\
&= \sum_{r=0}^{\alpha} \left(\mathcal{P}^r(h_0)\mathcal{P}^{\alpha-r}\mathcal{P}^{\Delta_0} + \cdots + \mathcal{P}^r(h_m)\mathcal{P}^{\alpha-r}\mathcal{P}^{\Delta_m}\right) \\
&= \sum_{r=0}^{\alpha} \left(\mathcal{P}^r(h_0)\mathcal{P}^{\Delta_0}\mathcal{P}^{\alpha-r} + \cdots + \mathcal{P}^r(h_m)\mathcal{P}^{\Delta_m}\mathcal{P}^{\alpha-r}\right) \\
&\quad + \sum_{r=0}^{\alpha} (\alpha - r)\mathcal{P}^r(h_0)\mathcal{P}^{\alpha-r} \\
&\quad + \sum_{r=0}^{\alpha} \left(\mathcal{P}^r(h_1)\mathcal{P}^{\Delta_2}\mathcal{P}^{\alpha-r-q} + \cdots + \mathcal{P}^r(h_m)\mathcal{P}^{\Delta_{m+1}}\mathcal{P}^{\alpha-r-q^m}\right) \\
&= \sum_{r=0}^{\alpha} \left(\mathcal{P}^r(h_0)\mathcal{P}^{\Delta_0}\mathcal{P}^{\alpha-r} + \cdots + \mathcal{P}^r(h_m)\mathcal{P}^{\Delta_m}\mathcal{P}^{\alpha-r}\right) \\
&\quad + \sum_{r=0}^{\alpha} \left((\alpha - r)\mathcal{P}^r(h_0)\mathcal{P}^{\alpha-r}\right) + \sum_{r=0}^{\alpha-q} \left(\mathcal{P}^r(h_1)\mathcal{P}^{\Delta_2}\mathcal{P}^{\alpha-r-q}\right) + \cdots \\
&\quad \cdots + \sum_{r=0}^{\alpha-q^m} \left(\mathcal{P}^r(h_m)\mathcal{P}^{\Delta_{m+1}}\mathcal{P}^{\alpha-r-q^m}\right) \\
&= \sum_{r=0}^{\alpha} \left(\mathcal{P}^r(h_0)\mathcal{P}^{\Delta_0}\mathcal{P}^{\alpha-r} + \cdots + \mathcal{P}^r(h_m)\mathcal{P}^{\Delta_m}\mathcal{P}^{\alpha-r}\right) \\
&\quad + \sum_{r=1}^{\alpha} \left(\mathcal{P}^{r-1}(h_0)\mathcal{P}^1\mathcal{P}^{\alpha-r}\right) + \sum_{r=q}^{\alpha} \left(\mathcal{P}^{r-q}(h_1)\mathcal{P}^{\Delta_2}\mathcal{P}^{\alpha-r}\right) + \cdots \\
&\quad \cdots + \sum_{r=q^m}^{\alpha} \left(\mathcal{P}^{r-q^m}(h_m)\mathcal{P}^{\Delta_{m+1}}\mathcal{P}^{\alpha-r}\right)
\end{aligned}
$$

$$= \Delta_\alpha + \Delta_{\alpha-1}\mathcal{P}^1 + \cdots + \Delta_0\mathcal{P}^\alpha$$

where we made extensive use of the Cartan formulae, Lemma 1.1.2, and the Adem-Wu relation:

$$\mathcal{P}^1\mathcal{P}^{\alpha-r} = (\alpha - r + 1)\mathcal{P}^{\alpha-r+1}.$$

By the induction hypothesis $\Delta_0 = \cdots = \Delta_{\alpha-1} = 0$, hence

$$\Delta_\alpha = 0,$$

as we claimed. •

Chapter 2
Some Field Theory over the Steenrod Algebra

In this chapter we are going to set up the category of graded fields over the Steenrod algebra in a proper way. We also generalize to the case of general Galois fields another ingredient we will need to prove the Embedding Theorem, namely the Separable Extension Lemma due to Clarence W. Wilkerson, see [43] or [36] Lemma 10.5.3. After that we will have a look at the much more delicate case of inseparable field extensions. We will give a proper definition of inseparability and inseparably closed in our category, construct the inseparable closure of a field over $\mathcal{P}^*$, and prove some of its basic properties, and provide some illustrative examples.

2.1 Graded Fields over the Steenrod Algebra

In this section we collect the definitions needed to introduce the **category of graded fields over the Steenrod algebra** $\mathcal{P}^*$.

A **graded field** $\mathbb{K}^*$ of characteristic p is a graded connected commutative algebra over $\mathbb{F}_q$ without zero divisors

$$\mathbb{K}^* = \{\mathbb{K}^*_{(i)} \mid i \in \mathbb{Z}\},$$

where every homogeneous element is invertible. Note that there are also elements with negative degree. This definition does not imply that inhomogeneous elements of the totalization[1] need to be invertible. So, if we forget the grading and take the totalization, we have only a *ring*, i.e., there is a difference between the totalization of a graded field $\mathbb{K}^*$, which need not to be a field, and its field of fractions.

If $\mathbb{K}^* \subseteq \mathbb{L}^*$ is an inclusion of graded fields then we say $\mathbb{L}^*$ **is an extension of** $\mathbb{K}^*$ and denote this by $\mathbb{L}^*/\mathbb{K}^*$. Let $\mathbb{L}^*/\mathbb{K}^*$ be an extension of graded fields, and

[1] By the totalization we understand the ring $\mathrm{tot}(\mathbb{K}^*) = \oplus_i \mathbb{K}^*_{(i)}$, compare [36] §4.1.

consider a polynomial

$$f(X) \in \mathbb{K}^*[X]$$

in one variable with coefficients in $\mathbb{K}^*$. Let $f(X)$ have degree d as polynomial in X, i.e.,

$$f(X) = k_d X^d + k_{d-1} X^{d-1} + \cdots + k_1 X + k_0,$$

with $k_d \neq 0$. Obviously,

$$f(l) = k_d l^d + \cdots + k_1 l + k_0 \notin \mathbb{L}^*$$

for an arbitrary $l \in \mathbb{L}^*$, because this expression may not be homogeneous. This means evaluating a polynomial $f(X)$ at $l \in \mathbb{L}^*$ does not lead to an element of $\mathbb{L}^*$ unless

$$\deg(k_i) + i \deg(l)$$

is a constant for all $i = 0, \ldots, d$. Call a polynomial that satisfies this condition $\deg(l)$-**graded**. For such a polynomial we can speak of its roots in $\mathbb{L}^*$ of degree c where $\deg(k_i) + ic$ is a constant. Hence the following definitions[2] make sense.

DEFINITION: *Let $\mathbb{K}^*$ be a graded field. An irreducible polynomial over $\mathbb{K}^*$, $f(X) \in \mathbb{K}^*[X]$, is called* ***-separable**, *resp.* ***-inseparable**, *if the derivative*

$$\frac{\mathrm{d}}{\mathrm{d}X} f(X)$$

does not vanish, resp. vanishes. An arbitrary polynomial is called ***-separable** *if all irreducible factors are separable. Otherwise call it* ***-inseparable**. *Let $\mathbb{L}^*/\mathbb{K}^*$ be an extension of graded fields, and $l \in \mathbb{L}^*$. A monic irreducible polynomial in $\mathbb{K}^*[X]$ is called* ***-minimal polynomial of** l, *written*

$$\mathrm{minpol}_{l \in \mathbb{L}^*/\mathbb{K}^*}(X) \in \mathbb{K}^*[X]$$

if $\mathrm{minpol}_{l \in \mathbb{L}^*/\mathbb{K}^*}(X)$ is $\deg(l)$-graded and

$$\mathrm{minpol}_{l \in \mathbb{L}^*/\mathbb{K}^*}(l) = 0 \in \mathbb{L}^*.$$

(Note that as in the ungraded case, $\mathrm{minpol}_{l \in \mathbb{L}^*/\mathbb{K}^*}(X)$ is the unique monic polynomial of minimal degree in X with root l.) We say that $l \in \mathbb{L}^*/\mathbb{K}^*$ is ***-algebraic** *over $\mathbb{K}^*$ if there exists a *-minimal polynomial for l. We call l* ***-separable**, *resp.* ***-inseparable**, *if its *-minimal polynomial is *-separable, resp. *-inseparable. If l has no *-minimal polynomial, we call it* ***-transcendental**. *Call the field extension $\mathbb{L}^*/\mathbb{K}^*$* ***-algebraic**, *if every $l \in \mathbb{L}^*/\mathbb{K}^*$ is *-algebraic over $\mathbb{K}^*$; otherwise call it a* ***-transcendental** *field extension. If we obtain $\mathbb{L}^*$ from $\mathbb{K}^*$ by adjoining elements of some trancendence set over $\mathbb{K}^*$, then the $\mathbb{L}^*/\mathbb{K}^*$ is called a* ***-purely trancendental** *field extension. A *-algebraic extension is* ***-separable** *if every $l \in \mathbb{L}^*/\mathbb{K}^*$ is *-separable; otherwise call it an* ***-inseparable** *field extension.*

[2] Compare [48], Chapter II, Section 12, for the classical terminology.

For a polynomial $f(X) \in \mathbb{K}^*[X]$ write[3]

$$f(X) = \tilde{f}(X^{p^e})$$

for some $e \geq 0$, where e is maximal with respect to this property (and, of course, p is the characteristic of $\mathbb{K}^*$), i.e.,

$$f(X) \in \mathbb{K}^*[X^{p^e}], \quad \text{but} \quad f(X) \notin \mathbb{K}^*[X^{p^{e+1}}].$$

Then[4]

$$\deg(f) = \deg(\tilde{f})p^e,$$

where p does not divide $\deg(\tilde{f})$. The integer $\deg(\tilde{f})$ is called the **degree of separability** of f, while p^e is the **degree of inseparability** of f (and e is the **exponent of inseparability**). So, a polynomial $f(X) \in \mathbb{K}^*[X]$ is *-separable if and only if

$$f(X) = \tilde{f}(X), \quad \text{and} \quad e = 0.$$

DEFINITION: *An element $l \in \mathbb{L}^*/\mathbb{K}^*$ is called* ***-purely inseparable** *over $\mathbb{K}^*$, if its *-minimal polynomial is*

$$X^{p^e} - l$$

*for some $e \geq 0$, or, equivalently, if the degree of separability of its *-minimal polynomial is one. Thus call a field extension $\mathbb{L}^*/\mathbb{K}^*$* ***-purely inseparable** *if every element $l \in \mathbb{L}^*$ is *-purely inseparable over $\mathbb{K}^*$.*

As expected we get that a field extension $\mathbb{L}^*/\mathbb{K}^*$ is *-separable and at the same time *-purely inseparable if and only if $\mathbb{L}^* = \mathbb{K}^*$.

We can talk about **graded fields $\mathbb{K}^*$ over the Steenrod algebra $\mathcal{P}^*$** (or **$\mathcal{P}^*$-graded fields** for short): just replace the word "algebra" by "field" in the definition in the introduction. Note carefully that the second part of the unstability condition doesn't make sense anymore, since for a non-zero element $0 \neq k \in \mathbb{K}^*_{(n)}$ of positive degree n we have $k^{-1} \in \mathbb{K}^*_{(-n)}$. So, the second part of the unstability condition would imply that $k^{-1} = \mathcal{P}^0(k^{-1}) = 0$, which is nonsense. In other words: for graded fields unstability is not meaningful.

DEFINITION: *Let $\mathbb{K}^*$ and $\mathbb{L}^*$ be fields over the Steenrod algebra, and let $\mathbb{K}^* \subseteq \mathbb{L}^*$. If the Steenrod algebra action on $\mathbb{L}^*$ restricted to $\mathbb{K}^*$ coincides with that one on $\mathbb{K}^*$ (i.e., if the inclusion commutes with the Steenrod algebra action), we say that $\mathbb{L}^*/\mathbb{K}^*$ is a* **field extension over the Steenrod algebra**[5] *(i.e., the inclusion is a morphism in the category of fields, or more generally algebras, over the Steenrod algebra). Call a $\mathcal{P}^*$-field extension* **$\mathcal{P}^*$-algebraic,**

[3] See [48], Chapter II, Section 5, for the classical case.

[4] $\deg(f)$ is the degree of f in X.

[5] $\mathcal{P}^*$-field extension, for short.

resp. $\mathcal{P}^*$**-transcendental**, *if it is an *-algebraic, resp. *-transcendental, extension. Call the field extension* $\mathcal{P}^*$**-separable**, $\mathcal{P}^*$**-inseparable**, *resp.* $\mathcal{P}^*$**-purely inseparable**, *if it is *-separable, *-inseparable, resp. *-purely inseparable.*

2.2 Separable Extensions

In this section we prove the general version of the Separable Extension Lemma, Proposition 2.2 in [43]. To wit:

Consider a $*$-separable field extension $\mathbb{L}^*/\mathbb{K}^*$, where the smaller field $\mathbb{K}^*$ enjoys an action of the Steenrod algebra $\mathcal{P}^*$. We are going to extend this action uniquely to a $\mathcal{P}^*$-action on $\mathbb{L}^*$. Take an arbitrary element $l \in \mathbb{L}^*$. It has a $*$-separable minimal polynomial

$$\mathrm{minpol}_{l\in\mathbb{L}^*/\mathbb{K}^*}(X) = \sum_{j=0}^{m} \kappa_j X^j \in \mathbb{K}^*[X]$$

of degree m over $\mathbb{K}^*$. We assume without loss of generality that $m > 1$, for otherwise we had $l \in \mathbb{K}^*$ and nothing to do. Consider the simple field extension

$$\mathbb{K}^* \hookrightarrow \mathbb{K}^*(l).$$

We have

LEMMA 2.2.1: *If* $\mathbb{L}^* = \mathbb{K}^*(l)/\mathbb{K}^*$ *is a simple $*$-separable field extension, and if* $\mathbb{K}^*$ *carries an action of the Steenrod algebra* $\mathcal{P}^*$, *then this action can be uniquely extended to* $\mathbb{K}^*(l)$.

PROOF: First we define $\mathcal{P}^i(l^j)$ by double induction on $i \geq 0$ and $j \geq 1$.

If $i = 0$, set

$$\mathcal{P}^0(l^j) = l^j \quad \forall\, j \geq 1.$$

If $i \geq 1$, assume that $\mathcal{P}^k(l^j)$ has been defined for all $k < i$ and all $j \geq 1$. We employ the minimal polynomial of l to define $\mathcal{P}^i(l)$. Formally, since $0 \in \mathbb{K}^*$, we have

$$\begin{aligned}
0 = \mathcal{P}^i_{\mathbb{K}^*}\left(\sum_{j=0}^{m} \kappa_j l^j\right) &= \sum_{j=0}^{m}\left(\sum_{r+s=i} \mathcal{P}^r_{\mathbb{K}^*}(\kappa_j)\mathcal{P}^s_{\mathbb{L}^*}(l^j)\right)\\
&= \sum_{j=0}^{m}\left(\sum_{r+s=i,\ s<i} \mathcal{P}^r_{\mathbb{K}^*}(\kappa_j)\mathcal{P}^s_{\mathbb{L}^*}(l^j)\right) + \sum_{j=1}^{m} \kappa_j \mathcal{P}^i_{\mathbb{L}^*}(l^j)\\
&= \sum_{j=0}^{m}\left(\sum_{r+s=i,\ s<i} \mathcal{P}^r_{\mathbb{K}^*}(\kappa_j)\mathcal{P}^s_{\mathbb{L}^*}(l^j)\right)\\
&\quad + \sum_{j=1}^{m} \kappa_j \left(\sum_{r_1+\cdots+r_j=i,\ r_1,\ldots,r_j<i} \mathcal{P}^{r_1}_{\mathbb{L}^*}(l)\cdots\mathcal{P}^{r_j}_{\mathbb{L}^*}(l)\right) + \sum_{j=1}^{m} j\kappa_j \mathcal{P}^i_{\mathbb{L}^*}(l)l^{j-1}.
\end{aligned}$$

Since $\mathrm{minpol}_{l\in\mathbb{L}^*/\mathbb{K}^*}(X)$ is *-separable we have that

$$0 \neq \frac{\mathrm{d}}{\mathrm{d}X}\mathrm{minpol}_{l\in\mathbb{L}^*/\mathbb{K}^*}(X)\Big|_{X=l} = \sum_{j=1}^{m} j\kappa_j l^{j-1}.$$

Hence we can solve the equation formally for $\mathcal{P}^i_{\mathbb{L}^*}(l)$ (remember, that the $\mathcal{P}^s_{\mathbb{L}^*}(l)$ are defined for $s < i$) and define

$$\mathcal{P}^i_{\mathbb{L}^*}(l) :=$$

$$-\frac{\sum\limits_{j=0}^{m}\left(\sum\limits_{r+s=i,\ s<i}\mathcal{P}^r_{\mathbb{K}^*}(\kappa_j)\mathcal{P}^s_{\mathbb{L}^*}(l^j)\right)+\sum\limits_{j=1}^{m}\kappa_j\left(\sum\limits_{r_1+\cdots+r_j=i,\ r_1,\ldots,\ r_j<i}\mathcal{P}^{r_1}_{\mathbb{L}^*}(l)\cdots\mathcal{P}^{r_j}_{\mathbb{L}^*}(l)\right)}{\sum\limits_{j=1}^{m} j\kappa_j l^{j-1}}.$$

Then, we can define inductively for all $j \geq 1$

$$\mathcal{P}^i_{\mathbb{L}^*}(l^j) = \sum_{r=0}^{i}\mathcal{P}^r_{\mathbb{L}^*}(l)\mathcal{P}^{i-r}_{\mathbb{L}^*}(l^{j-1}).$$

$\mathbb{K}^*(l)$ is generated by $1,\ l,\ldots,\ l^{m-1}$ over $\mathbb{K}^*$, so the following formula tells us how to evaluate $\mathcal{P}^i_{\mathbb{L}^*}$ on an arbitrary element of $\mathbb{K}^*(l)$.

$$\mathcal{P}^i_{\mathbb{L}^*}\left(\sum_{j=0}^{m-1}k_j l^j\right) := \sum_{j=0}^{m-1}\mathcal{P}^i_{\mathbb{L}^*}\left(k_j l^j\right) := \sum_{j=0}^{m-1}\sum_{r+s=i}\left(\mathcal{P}^r_{\mathbb{K}^*}(k_j)\mathcal{P}^s_{\mathbb{L}^*}(l^j)\right) \quad \forall\ k_j \in \mathbb{K}^*.$$

This action of the Steenrod powers $\mathcal{P}^i_{\mathbb{L}^*}$ on $\mathbb{K}^*(l)$ is by construction additive. It also satisfies the Cartan formulae as one sees as follows. Define a ring homomorphism

$$\begin{array}{rccc} F(\xi): & \mathbb{K}^*[X] & \longrightarrow & \mathbb{K}^*(l)[[\xi]] \\ & \sum\limits_{j=0}^{m-1} k_j X^j & \longmapsto & \mathbf{P}(\xi)\left(\sum\limits_{j=0}^{m-1} k_j l^j\right) = \sum\limits_{j=0}^{m-1}\mathbf{P}(\xi)(k_j)\mathbf{P}(\xi)(l)^j, \end{array}$$

where $\mathbf{P}(\xi)$ denotes the giant Steenrod operation. We have that an element $\sum\limits_{j=0}^{m'} k_j X^j \in \mathbb{K}^*[X]$ is in the kernel of this map if and only if

$$\mathbf{P}(\xi)\left(\sum_{j=0}^{m'} k_j l^j\right) = 0.$$

This is equivalent to saying that all coefficients in this expression vanish, i.e.,

$$\mathcal{P}^i_{\mathbb{L}^*}\left(\sum_{j=0}^{m'} k_j l^j\right) = 0 \quad \forall\ i \geq 0$$

which in turn is equivalent to

$$\sum_{j=0}^{m'} k_j l^j = 0.$$

Therefore the kernel of $F(\xi)$ is generated by the minimal polynomial of l, and, $F(\xi)$ factorizes through the map

$$\boldsymbol{P}(\xi) : \mathbb{K}^*[X]/\left(\sum_{j=0}^{m} \kappa_j X^j\right) \cong \mathbb{K}^*(l) \longrightarrow \mathbb{K}^*(l)[[\xi]],$$

what means that the Cartan formulae are valid.

So far we have an action of the free associative $\mathbb{F}$-algebra generated by the $\mathcal{P}^i$'s extended to $\mathbb{K}^*(l)$, and satisfying the Cartan formulae. It remains to show that the Adem-Wu relations are valid. We employ the Bullett-Macdonald identity to do so. Recalling the notation from the introduction we have

$$\begin{aligned}
0 &= \left(\boldsymbol{P}(s)\boldsymbol{P}(1) - \boldsymbol{P}(u)\boldsymbol{P}(t^q)\right)\left(\sum_{j=0}^{m} \kappa_j l^j\right) \\
&= \sum_{j=0}^{m} \left(\boldsymbol{P}(s)\boldsymbol{P}(1) - \boldsymbol{P}(u)\boldsymbol{P}(t^q)\right)\left(\kappa_j l^j\right) \\
&= \sum_{j=0}^{m} \left(\boldsymbol{P}(s)\boldsymbol{P}(1) - \boldsymbol{P}(u)\boldsymbol{P}(t^q)\right)\left(\kappa_j\right)\left(\boldsymbol{P}(s)\boldsymbol{P}(1) - \boldsymbol{P}(u)\boldsymbol{P}(t^q)\right)(l)^j \\
&= \left(\boldsymbol{P}(s)\boldsymbol{P}(1) - \boldsymbol{P}(u)\boldsymbol{P}(t^q)\right)(l)^j
\end{aligned}$$

where the last equation follows, because the Bullett-Macdonald identity evaluates to zero on $\mathbb{K}^*$ by assumption. Since $\mathbb{K}^*(l)$ is a field and has in particular no nilpotent elements it follows that

$$\boldsymbol{P}(s)\boldsymbol{P}(1)(l) - \boldsymbol{P}(u)\boldsymbol{P}(t^q)(l) = 0.$$

Therefore $\mathbb{K}^*(l)$ carries an action of the Steenrod algebra.

Finally we show that this action of the Steenrod algebra on $\mathbb{K}^*(l)$ is the unique action extending the given action on $\mathbb{K}^*$. It suffices to check this for the reduced power operations $\mathcal{P}^i$. For $i = 0$ there is nothing to show. Let $i > 0$, and assume there were another extension $\overline{\mathcal{P}^i_{\mathbb{L}^*}}$. Since $\mathbb{K}^*(l)/\mathbb{K}^*$ is $*$-separable, we have

$$\mathbb{K}^*(l) = \mathbb{K}^*(l^p) = \mathbb{K}^*(l^q),$$

by [48] the Corollary on page 70. Therefore $l \in \mathbb{K}^*(l^q)$, i.e., there exist elements $k_0, \ldots, k_m \in \mathbb{K}^*$ such that

$$l = \sum_{j=1}^{m} k_j (l^q)^j + k_0.$$

Then, we get by induction

$$\begin{aligned}\overline{\mathcal{P}^i_{\mathbb{L}^*}}(l)-\mathcal{P}^i_{\mathbb{L}^*}(l) &= \sum_{j=1}^{m}\left(\sum_{r+s=i}\mathcal{P}^r_{\mathbb{K}^*}(k_j)\left(\overline{\mathcal{P}^s_{\mathbb{L}^*}}(l^{qj})-\mathcal{P}^s_{\mathbb{L}^*}(l^{qj})\right)\right)\\ &= \sum_{j=1}^{m}k_j\left(\overline{\mathcal{P}^i_{\mathbb{L}^*}}(l^{qj})-\mathcal{P}^i_{\mathbb{L}^*}(l^{qj})\right)\\ &= \begin{cases}\sum_{j=1}^{m}k_j\left(\overline{\mathcal{P}^{\frac{i}{q}}_{\mathbb{L}^*}}(l^j)-\mathcal{P}^{\frac{i}{q}}_{\mathbb{L}^*}(l^j)\right)^q & \text{for } q\mid i\\ 0 & \text{otherwise}\end{cases}\\ &= 0,\end{aligned}$$

where we have used the induction hypothesis twice. •

The general version is now an application[6] of Zorn's Lemma.

PROPOSITION 2.2.2 (Separable Extension Lemma, General Version): *Let $\mathbb{L}^*/\mathbb{K}^*$ be a *-separable extension of a field $\mathbb{K}^*$ of characteristic p, and suppose that $\mathbb{K}^*$ is a graded field over the Steenrod algebra $\mathcal{P}^*$. Then there exists a unique extension* [7] *of this action to $\mathbb{L}^*$.*

PROOF: We consider the set $\mathcal{S}$ of intermediate fields Δ^*

$$\mathbb{K}^* \hookrightarrow \Delta^* \hookrightarrow \mathbb{L}^*,$$

which carry an action of the Steenrod algebra. Since $\mathbb{K}^* \in \mathcal{S}$, the set is not empty. Moreover $\mathcal{S}$ has maximal elements, since every ascending chain of fields in $\mathcal{S}$ has an upper bound. So choose a maximal element $\Delta^*_{\max} \in \mathcal{S}$. If this were strictly smaller than $\mathbb{L}^*$, there would be an element $l \in \mathbb{L}^*$ which is not in $\Delta^*_{\max}$. Since l is *-separable over $\mathbb{K}^*$, it is a fortiori *-separable over $\Delta^*_{\max}$. Hence

$$\Delta^*_{\max} \subsetneqq \Delta^*_{\max}(l).$$

However, Lemma 2.2.1 tells us that we can extend the $\mathcal{P}^*$-action from $\Delta^*_{\max}$ to $\Delta^*_{\max}(l)$ contrary to the maximality of $\Delta^*_{\max}$ in $\mathcal{S}$. •

This result tells us that the notion of *-separability and $\mathcal{P}^*$-separability are equivalent as long as we have an action of the Steenrod algebra on the smaller field. We come to the much more delicate case of inseparable field extensions.

---

[6] Of course, for finite field extensions a simple induction argument would suffice.

[7] In other words, a *-separable field extension, where the little field admits an action of the Steenrod algebra, is automatically a separable field extension in the category of fields over the Steenrod algebra, i.e., is a $\mathcal{P}^*$-separable field extension. However, be prepared! The notion of *-inseparability can't be translated that easily to fields over the Steenrod algebra.

2.3 Inseparable Extensions

In this section we ask ourselves the obvious question: is there a "purely inseparable extension lemma", i.e., whether every *-inseparable field extension $\mathbb{L}^*/\mathbb{K}^*$ of a field $\mathbb{K}^*$ over the Steenrod algebra, is $\mathcal{P}^*$-inseparable.

Let us first have a look at an example:

Let $\mathbb{K}^* \subseteq \mathbb{L}^*$ be a field extension of graded fields. Let $l \in \mathbb{L}^*$ be a *-purely inseparable element over $\mathbb{K}^*$ with *-minimal polynomial

$$\mathrm{minpol}_{l \in \mathbb{L}^*/\mathbb{K}^*}(X) = X^{p^e} - \kappa \in \mathbb{K}^*[X].$$

Then

$$\deg(l)p^e = \deg(\kappa),$$

because the minimal polynomial has to be $\deg(l)$-graded (see definitions in Section 2.1).

This is not enough! Have a look at the field $\mathbb{F}_2(x, y)$ generated by two transcendental elements of degree 1 over the field of two elements. Consider the following *-purely inseparable polynomial

$$f(X) := X^2 - xy \in \mathbb{F}_2(x, y)[X].$$

Its splitting field $\mathbb{L}^*$ would contain an element l of degree one such that

$$l^2 = xy \in \mathbb{F}_2(x, y).$$

This splitting field is *not* an object in our category, for otherwise, there would exist an extension of the (naturally given) Steenrod algebra action on $\mathbb{F}_2(x, y)$ to $\mathbb{L}^*$. However,

$$x^2y + xy^2 = \mathcal{P}^1(xy) = \mathcal{P}^1(l^2) = 2\mathcal{P}^1(l)l = 0$$

shows that this can't be done consistently. So, what we need is that the p-th powers in the little field "behave as such" under the action of the Steenrod algebra. We achieve this by taking into account the additional derivations which are given to us by the Steenrod algebra, namely $\mathcal{P}^{\Delta_i}$ for $i \geq 0$ (recall Section 1.1):

DEFINITION: *Let $\mathbb{K}^*$ be a field over the Steenrod algebra. A polynomial $f(X) \in \mathbb{K}^*[X]$ is called* **purely inseparable over $\mathcal{P}^*$**, *or* **$\mathcal{P}^*$-purely inseparable** *if*

(1) *it has the form*

$$f(X) = X^{p^e} - \kappa$$

for some $e \in \mathbb{N}_0$ and $\kappa \in \mathbb{K}^*$, and

(2) *if $e \geq 1$ then in addition*

$$\mathcal{P}^{\Delta_i}(\kappa) = 0 \quad \forall\, i \geq 0.$$

The following theorem justifies this terminology.

THEOREM 2.3.1: *Let $\mathbb{L}^*/\mathbb{K}^*$ be an extension of fields over the Steenrod algebra. Then it is $\mathcal{P}^*$-purely inseparable if and only if the minimal polynomial of every element $l \in \mathbb{L}^*$ is $\mathcal{P}^*$-purely inseparable.*

PROOF: The "if"-case is trivial. So, consider that $\mathbb{L}^*/\mathbb{K}^*$ is a field extension over the Steenrod algebra. If it is $\mathcal{P}^*$-purely inseparable then every element $l \in \mathbb{L}^*$ has a minimal polynomial

$$X^{p^e} - \kappa \in \mathbb{K}^*[X].$$

If $e = 0$ this polynomial is $\mathcal{P}^*$-purely inseparable by definition. If $e \geq 1$, then since $l^{p^e} = \kappa$ and the actions of the Steenrod algebra on $\mathbb{K}^*$ and $\mathbb{L}^*$ are compatible we have

$$\mathcal{P}^{\Delta_i}(\kappa) = \mathcal{P}^{\Delta_i}(l^{p^e}) = 0$$

as claimed. •

DEFINITION: *A field $\mathbb{K}^*$ over the Steenrod algebra is $\mathcal{P}^*$-inseparably closed if any $\mathcal{P}^*$-algebraic field extension $\mathbb{L}^*/\mathbb{K}^*$ is $\mathcal{P}^*$-separable.*

Every graded field over the Steenrod algebra has a **$\mathcal{P}^*$-inseparable closure**, which can be obtained in the following way. First we take the algebraic closure of $\mathbb{K}^*$ as a graded field, $\overline{\mathbb{K}^*}$. This exists and the classical non-graded proof works more or less word-for-word, see e.g. [48] Volume 1, Chapter II, Section 14. Inside $\overline{\mathbb{K}^*}$ we can take the $*$-inseparable closure[8] $\mathbb{K}^{p^{-\infty}}$ of $\mathbb{K}^*$ as a graded field. Hence we get a chain of graded fields

$$\mathbb{K}^* \hookrightarrow \mathbb{K}^{p^{-\infty}} \hookrightarrow \overline{\mathbb{K}^*},$$

where the first extension, $\mathbb{K}^* \hookrightarrow \mathbb{K}^{p^{-\infty}}$, is $*$-purely inseparable, while the second is $*$-separable. We get the $\mathcal{P}^*$-inseparable closure, $\mathbb{K}^*{}_{\mathcal{P}^*\text{-insep}}$, of the field $\mathbb{K}^*$ in the *category of graded fields over the Steenrod algebra* by taking the graded subfield of $\mathbb{K}^{p^{-\infty}}$ generated by the roots of $\mathcal{P}^*$-purely inseparable polynomials with coefficients in $\mathbb{K}^*$:

$$\mathbb{K}^* \hookrightarrow \mathbb{K}^*{}_{\mathcal{P}^*\text{-insep}} \hookrightarrow \mathbb{K}^{p^{-\infty}} \hookrightarrow \overline{\mathbb{K}^*}.$$

Specifically, we construct the $\mathcal{P}^*$-inseparable closure $\mathbb{K}^*{}_{\mathcal{P}^*\text{-insep}}$ by the following inductive procedure. Let $\mathbb{K}^* = \mathbb{K}^*{}_0$ and define successively

$$\mathbb{K}^*{}_i = \mathbb{K}^*{}_{i-1}(\mathcal{S})$$

where the set $\mathcal{S}$ consists of the roots l of $\mathcal{P}^*$-purely inseparable polynomials of degree p in X, i.e., $l^p \in \mathbb{K}^*{}_{i-1}$ and $\mathcal{P}^{\Delta_i}(l^p) = 0$ for all $i \geq 0$. In order to extend the Steenrod action to $\mathbb{K}^*{}_i$ in consistency with the Cartan formulae we set

$$\mathcal{P}^j_{\mathbb{K}^*{}_{i-1}}(l^p) = \mathcal{P}^j_{\mathbb{K}^*{}_i}(l^p) = \begin{cases} \left(\mathcal{P}^k_{\mathbb{K}^*{}_i}(l)\right)^p & \text{for } kp = j \\ 0 & \text{otherwise.} \end{cases}$$

[8] We adopt the notation from [48], page 108.

Since $\mathbb{K}^*_i$ contains *all* roots of $\mathcal{P}^*$-purely inseparable polynomials of degree p it contains also the root of

$$X^p - \left(\mathcal{P}^k_{\mathbb{K}^*_i}(l)\right)^p \in \mathbb{K}^*_{i-1}[X],$$

i.e., $\mathbb{K}^*_i$ is closed under the so-defined action of the $\mathcal{P}^j$'s. Since

$$\begin{aligned}\left(\mathcal{P}^k_{\mathbb{K}^*_i}(l) + \mathcal{P}^k_{\mathbb{K}^*_i}(l')\right)^p &= \mathcal{P}^{kp}_{\mathbb{K}^*_{i-1}}(l^p) + \mathcal{P}^{kp}_{\mathbb{K}^*_{i-1}}(l'^p)\\ &= \mathcal{P}^{kp}_{\mathbb{K}^*_{i-1}}(l^p + l'^p)\\ &= \mathcal{P}^{kp}_{\mathbb{K}^*_{i-1}}\left((l + l')^p\right)\\ &= \left(\mathcal{P}^k_{\mathbb{K}^*_i}(l + l')\right)^p,\end{aligned}$$

for all $l,\ l' \in \mathbb{K}^*_i$ and we don't have any nilpotent elements this action is additive. The validity of the Cartan formulae can be shown in a similar[9] manner.

$$\begin{aligned}\left(\mathcal{P}^k_{\mathbb{K}^*_i}(ll')\right)^p &= \mathcal{P}^{kp}_{\mathbb{K}^*_{i-1}}(l^p l'^p) = \sum_{\alpha+\beta=kp} \mathcal{P}^{\alpha}_{\mathbb{K}^*_{i-1}}(l^p)\mathcal{P}^{\beta}_{\mathbb{K}^*_{i-1}}(l'^p)\\ &= \left(\sum_{\alpha+\beta=kp} \mathcal{P}^{\frac{\alpha}{p}}_{\mathbb{K}^*_i}(l)\mathcal{P}^{\frac{\beta}{p}}_{\mathbb{K}^*_i}(l')\right)^p = \left(\sum_{\alpha'+\beta'=k} \mathcal{P}^{\alpha'}_{\mathbb{K}^*_i}(l)\mathcal{P}^{\beta'}_{\mathbb{K}^*_i}(l')\right)^p\end{aligned}$$

for all $l,\ l' \in \mathbb{K}^*_i$ and therefore

$$\mathcal{P}^k_{\mathbb{K}^*_i}(ll') = \sum_{\alpha'+\beta'=k} \mathcal{P}^{\alpha'}_{\mathbb{K}^*_i}(l)\mathcal{P}^{\beta'}_{\mathbb{K}^*_i}(l').$$

So we have an action of the free associative $\mathbb{F}$-algebra generated by the reduced power operations on $\mathbb{K}^*_i$. Next we prove that also the Adem-Wu relations are valid. We employ the easy to handle Bullett-Macdonald identity to do so. Recalling the notation from the introduction we have

$$\begin{aligned}0 &= \boldsymbol{P}(s)\boldsymbol{P}(1)(l^p) - \boldsymbol{P}(u)\boldsymbol{P}(t^q)(l^p)\\ &= \boldsymbol{P}(s)\left(\boldsymbol{P}(1)(l)\right)^p - \boldsymbol{P}(u)\left(\boldsymbol{P}(t^q)(l)\right)^p\\ &= \left(\boldsymbol{P}(s)\boldsymbol{P}(1)(l)\right)^p - \left(\boldsymbol{P}(u)\boldsymbol{P}(t^q)(l)\right)^p\\ &= \left(\boldsymbol{P}(s)\boldsymbol{P}(1)(l) - \boldsymbol{P}(u)\boldsymbol{P}(t^q)(l)\right)^p \quad \forall\, l \in \mathbb{K}^*_i.\end{aligned}$$

[9] Let $k \in \mathbb{K}^*_{i-1}$ be a $\mathcal{P}^{\Delta}$-constant. Then Lemma 1.1.2 gives inductively

$$\mathcal{P}^{\Delta_j}\mathcal{P}^m(k) = -\mathcal{P}^{\Delta_{j+1}}\mathcal{P}^{m-q^j}(k) = 0 \quad \forall\, j \geq 1.$$

Since $\mathcal{P}^{\Delta_0}\mathcal{P}^m(k) = \left(\deg(k) + m(q-1)\right)k$, it follows that $\mathcal{P}^m(k)$ is again a $\mathcal{P}^{\Delta}$-constant, if $m \equiv 0$ MOD (p). On the other hand, if m is not divisible by p, then the Adem-Wu relations give

$$m\mathcal{P}^m(k) = \mathcal{P}^1\mathcal{P}^{m-1}(k) = \mathcal{P}^{\Delta_1}\mathcal{P}^{m-1}(k) = 0,$$

where the last equation is valid by induction. Therefore $\mathcal{P}^m(k) = 0$ for all $\mathcal{P}^{\Delta}$-constants and whenever m is not divisible by p, so that the penultimate of the following equations makes sense. Compare Lemma 4.1.2.

Since $\mathbb{K}^*{}_i$ is a field and has, in particular, no nilpotent elements it follows that

$$\boldsymbol{P}(s)\boldsymbol{P}(1)(l) - \boldsymbol{P}(u)\boldsymbol{P}(t^q)(l) = 0$$

for every $l \in \mathbb{K}^*{}_i$. Therefore $\mathbb{K}^*{}_i$ carries an action of the Steenrod algebra.

This action is also unique as we next show. Let $l \in \mathbb{K}^*{}_i$ and $l^p \in \mathbb{K}^*{}_{i-1}$. Assume there were two extensions $\mathcal{P}^j_{\mathbb{K}^*{}_i}$ and $\overline{\mathcal{P}^j_{\mathbb{K}^*{}_i}}$. If $j = 0$, then

$$\mathcal{P}^0_{\mathbb{K}^*{}_i} = \mathrm{id}_{\mathbb{K}^*{}_i} = \overline{\mathcal{P}^0_{\mathbb{K}^*{}_i}}\,.$$

Let $j > 0$. Then we have

$$\begin{aligned}\left(\mathcal{P}^j_{\mathbb{K}^*{}_i}(l) - \overline{\mathcal{P}^j_{\mathbb{K}^*{}_i}(l)}\right)^p &= \left(\mathcal{P}^j_{\mathbb{K}^*{}_i}(l)\right)^p - \left(\overline{\mathcal{P}^j_{\mathbb{K}^*{}_i}(l)}\right)^p \\ &= \mathcal{P}^{jp}_{\mathbb{K}^*{}_i}(l^p) - \overline{\mathcal{P}^{jp}_{\mathbb{K}^*{}_i}}(l^p) \\ &= \mathcal{P}^{jp}_{\mathbb{K}^*{}_{i-1}}(l^p) - \overline{\mathcal{P}^{jp}_{\mathbb{K}^*{}_{i-1}}}(l^p) \\ &= 0,\end{aligned}$$

where the last equation holds since by induction the two actions coincide on $\mathbb{K}^*{}_{i-1}$. Hence

$$\mathcal{P}^j_{\mathbb{K}^*{}_i}(l) - \overline{\mathcal{P}^j_{\mathbb{K}^*{}_i}}(l) = 0,$$

what we claimed. We have therefore constructed an ascending chain of $\mathcal{P}^*$-purely inseparable field extensions

$$\mathbb{K}^* = \mathbb{K}^*{}_0 \subseteq \mathbb{K}^*{}_1 \subseteq \cdots \subseteq \mathbb{K}^*{}_i \subseteq \cdots.$$

PROPOSITION 2.3.2: *The colimit of the above chain of fields,* colim($\mathbb{K}^*{}_i$)*, is the $\mathcal{P}^*$-inseparable closure of $\mathbb{K}^*$ in the sense that*

(1) colim($\mathbb{K}^*{}_i$) *is $\mathcal{P}^*$-inseparably closed, and*

(2) *if $\mathbb{L}^*$ is a $\mathcal{P}^*$-inseparably closed field, containing $\mathbb{K}^*$, there exists an inclusion* colim($\mathbb{K}^*{}_i$) $\hookrightarrow \mathbb{L}^*$ *extending* $\mathbb{K}^* \hookrightarrow \mathbb{L}^*$.

PROOF: Since the filtered colimit of graded fields over $\mathcal{P}^*$ is a graded field over $\mathcal{P}^*$ and the extensions are all $\mathcal{P}^*$-purely inseparable we certainly get

$$\mathrm{colim}(\mathbb{K}^*{}_i) \subseteq \mathbb{K}^*{}_{\mathcal{P}^*\text{-insep}}.$$

On the other hand every element $k \in \mathbb{K}^*{}_{\mathcal{P}^*\text{-insep}}$ is the root of a $\mathcal{P}^*$-purely inseparable polynomial, i.e.,

$$k^{p^l} \in \mathbb{K}^* \quad \text{for large enough } l \in \mathbb{N}_0.$$

Hence $k \in \mathbb{K}^*{}_l \subseteq \mathrm{colim}(\mathbb{K}^*{}_i)$ and we are done. •

EXAMPLE 1: Take the field of two elements and adjoin two transcendental generators of degree 1, $\mathbb{F}_2(x, y)$, i.e., we are looking at the field of fractions of the ring of polynomials $\mathbb{F}_2[x, y]$ in two variables over the field of two elements. If we forget the grading and take the field of fractions of the totalization, then this is certainly not an inseparably closed field since, e.g.,

$$f_1(X) := X^2 + x$$

has no root in it. However, no root of f_1, say r, leads to a *homogeneous* equation

$$r^2 + x = 0,$$

because r would had to have degree $\frac{1}{2}$, i.e, this polynomial is not deg(r)-graded in the terminology of Section 2.1. Nevertheless, our field is also not *-inseparably closed, because, e.g., the polynomial

$$f_2(X) := X^2 + xy$$

has no root. In fact, in *our category*, where we take the $\mathcal{P}^*$-action into account, $\mathbb{F}_2(x, y)$ *is* $\mathcal{P}^*$-inseparably closed. To see this, assume the contrary, namely that $\mathbb{F}_2(x, y)$ were not $\mathcal{P}^*$-inseparably closed. Construct its $\mathcal{P}^*$-inseparable closure as above by taking the colimit of the ascending chain of $\mathcal{P}^*$-purely inseparable field extensions

$$\mathbb{F}_2(x, y) = \mathbb{K}^*_0 \subseteq \mathbb{K}^*_1 \subseteq \cdots \subseteq \mathbb{K}^*_i \subseteq \cdots.$$

In particular,

$$\mathbb{F}_2(x, y) = \mathbb{K}^*_0 \subsetneq \mathbb{K}^*_1,$$

hence there would be an $\mathcal{P}^*$-inseparable polynomial of degree 2

$$f_3(X) = X^2 + \lambda_1 x^2 + \lambda_2 xy + \lambda_3 y^2$$

with $\lambda_1, \lambda_2, \lambda_3 \in \mathbb{F}_2$ which does not split over $\mathbb{F}_2(x, y)$. Note however, $\mathcal{P}^*$-inseparability implies

$$\lambda_2 = 0,$$

because $0 = \mathcal{P}^{\Delta_i}(f_3) = \lambda_2(x^2y + xy^2)$. However, for every value of $\lambda_1, \lambda_3 \in \mathbb{F}_2$ the polynomial

$$X^2 + \lambda_1 x^2 + \lambda_3 y^2 = (X + \lambda_1 x + \lambda_3 y)^2$$

is not irreducible.

This example is not an accident as the next two results show.

THEOREM 2.3.3: *Let $\mathbb{K}^*$ be $\mathcal{P}^*$-inseparably closed, let t be $\mathcal{P}^*$-transcendental over $\mathbb{K}^*$ of degree 1. If the action of the Steenrod derivations on t is given by*[10]

$$\mathcal{P}^{\Delta_i}(t) = t^{q^i} \quad \forall\, i \geq 0$$

then $\mathbb{K}^*(t)$ is $\mathcal{P}^*$-inseparably closed.

[10] This condition means that under the operation of the $\mathcal{P}^{\Delta}$'s the element t "behaves" as a standard linear form.

PROOF: Consider a $\mathcal{P}^*$-purely inseparable field extension

$$\mathbb{K}^*(t) \hookrightarrow \mathbb{L}^*.$$

We have to show that $\mathbb{K}^*(t) = \mathbb{L}^*$. So, take an $l \in \mathbb{L}^*$. Its minimal polynomial is then $\mathcal{P}^*$-purely inseparable of, say, degree p^m:

$$\mathrm{minpol}_{l \in \mathbb{L}^*/\mathbb{K}^*(t)}(X) = X^{p^m} - \sum_{i=0}^{d} k_i t^i,$$

where $k_i \in \mathbb{K}^*\ \forall\ i = 0, \dots, d$, for some $d \in \mathbb{N}_0$. Hence

$$(\star) \qquad l^{p^m} - \sum_{i=0}^{d} k_i t^i = 0.$$

We need to show that $l \in \mathbb{K}^*(t)$, or what is the same thing $m = 0$. First, observe that

$$\deg(k_i) + i = p^m \deg(l)\ \forall\ i = 0, \dots, d,$$

by homogeneity of the equation $(\star)$. If we apply a Steenrod derivation $\mathcal{P}^{\Delta_j}$ to this equation, we get

$$\begin{aligned} 0 = \mathcal{P}^{\Delta_j}\left(l^{p^m} - \sum_{i=0}^{d} k_i t^i\right) &= \sum_{i=0}^{d}\left(\mathcal{P}^{\Delta_j}(k_i)t^i + k_i \mathcal{P}^{\Delta_j}(t^i)\right) \\ &= \sum_{i=0}^{d}\left(\mathcal{P}^{\Delta_j}(k_i)t^i + i k_i \mathcal{P}^{\Delta_j}(t) t^{i-1}\right) \\ &= \sum_{i=0}^{d}\left(\mathcal{P}^{\Delta_j}(k_i)t^i + i k_i t^{q^j} t^{i-1}\right) \\ &= \sum_{i=0}^{d}\left(\mathcal{P}^{\Delta_j}(k_i)t^i + i k_i t^{q^j+i-1}\right). \end{aligned}$$

Let's suppose that $m > 0$. We are going to show that the constant term occurring in $(\star)$ is itself a p-th power, i.e., that

$$\sum_{i=0}^{d} k_i t^i = k^p$$

for some $k \in \mathbb{K}^*(t)$, which would contradict the minimality of m, because then $(\star)$ becomes

$$0 = l^{p^m} - \sum_{i=0}^{d} k_i t^i = \left(l^{p^{m-1}} - k\right)^p,$$

and

$$0 = l^{p^{m-1}} - k.$$

We proceed by considering possible values of $j \geq 0$.

First take a j such that $q^j - 1 > d$. We get

$$\begin{aligned} 0 &= \sum_{i=0}^{d} \left(\mathcal{P}^{\Delta_j}(k_i) t^i + i k_i t^{q^j + i - 1} \right) \\ &= \left(\sum_{i=0}^{d} \mathcal{P}^{\Delta_j}(k_i) t^i \right) + \left(\sum_{i=q^j-1}^{q^j+d-1} (i - q^j + 1) k_{i-q^j+1} t^i \right), \end{aligned}$$

where we have arranged the sums by powers of t. Since t is transcendental over $\mathbb{K}^*$ we must have that all coefficients vanish; so the first sum gives

$$\mathcal{P}^{\Delta_j}(k_i) = 0 \ \forall\ i = 0, \ldots, d, \text{ and } q^j - 1 > d$$

while the second gives that

$$i k_i = 0 \quad \forall\ i = 0, \ldots, d,$$

i.e.,

$$k_i = 0 \ \forall\ i = 0, \ldots, d \mid i \not\equiv 0 \ \textbf{MOD}\ p.$$

So the higher Steenrod derivations, $\mathcal{P}^{\Delta_j}$ with j such that $q^j - 1 > d$, vanish on all coefficients k_i and moreover, the coefficients k_i, where i is not divisible by p, are all zero:

$$(\lozenge) \qquad l^{p^m} - \sum_{i=0}^{d} k_i t^i = l^{p^m} - \sum_{i=0}^{[\frac{d}{p}]} k_{ip} t^{ip},$$

and

$$\mathcal{P}^{\Delta_j}(k_{ip}) = 0 \quad \forall\, j \text{ with } q^j - 1 > d.$$

We next show that the lower degree Steenrod derivations $\mathcal{P}^{\Delta_0}, \ldots, \mathcal{P}^{\Delta_j}$ for j with $q^j - 1 \leq d$ vanish on the coefficients k_{ip}. To this end, note that we have for our non-zero coefficients

$$\deg(k_{ip}) = \deg(l) p^m - ip \equiv 0 \ \textbf{MOD}\ p,$$

so $\mathcal{P}^{\Delta_0}$ evaluates to zero on them. Next let $j > 0$, but still satisfy $q^j - 1 \leq d$, and apply $\mathcal{P}^{\Delta_j}$ to the equation $(\lozenge)$. We obtain

$$0 = \mathcal{P}^{\Delta_j} \left(l^{p^m} - \sum_{i=0}^{[\frac{d}{p}]} k_{ip} t^{ip} \right)$$

$$= \sum_{i=0}^{[\frac{d}{p}]} \left(\mathcal{P}^{\Delta_j}(k_{ip}) t^{ip} + ipk_{ip} t^{q^j+ip-1} \right)$$

$$= \sum_{i=0}^{[\frac{d}{p}]} \left(\mathcal{P}^{\Delta_j}(k_{ip}) t^{ip} \right),$$

whence all coefficients of the powers of t must vanish, because t is transcendental over $\mathbb{K}^*$, i.e.,

$$\mathcal{P}^{\Delta_j}(k_i) = 0 \ \forall \ i, \ j.$$

This in turn means that all the coefficients are p-th powers in $\mathbb{K}^*$ because $\mathbb{K}^*$ is $\mathcal{P}^*$-inseparably closed. Since raising to p-th powers is additive in characteristic p we have shown that our sum in (★)

$$l^{p^m} - \sum_{i=0}^{d} k_i t^i$$

is a p-th power and hence m was not minimal. This is a contradiction. •

The above theorem has no direct generalization to the case where the transcendental element t has higher degree. The following example illustrates this.

EXAMPLE 2: Let $\mathbb{F}(x, y)$ be the field generated by two linear transcendental elements over a Galois field $\mathbb{F} = \mathbb{F}_q$ with standard Steenrod algebra action. Consider the subfields

$$\mathbb{F}(x^{q-1}) \hookrightarrow \mathbb{F}(x^{q-1}, xy^q) \hookrightarrow \mathbb{F}(x, y).$$

An easy and straightforward calculation that shows that both subfields are again closed under the action of the Steenrod algebra. As one might equally easily convince oneself the smaller field, $\mathbb{F}(x^{q-1})$, is $\mathcal{P}^*$-inseparably closed.[11] Obviously the field extension

$$\mathbb{F}(x^{q-1}) \hookrightarrow \mathbb{F}(x^{q-1}, xy^q)$$

is $\mathcal{P}^*$-purely transcendental and the additional element xy^q has degree prime to the characteristic, but the extended field $\mathbb{F}(x^{q-1}, xy^q)$ is *not* $\mathcal{P}^*$-inseparably closed, because all Steenrod derivations vanish on the element

$$x^q y^q = x^{q-1} xy^q \in \mathbb{F}(x^{q-1}, xy^q),$$

but $xy \notin \mathbb{F}(x^{q-1}, xy^q)$.

COROLLARY 2.3.4: *The field of fractions of the polynomial functions in n variables over $\mathbb{F}$, $\mathbb{F}(x_1, \ldots, x_n)$, is $\mathcal{P}^*$-inseparably closed.*

---

[11] This extends easily to the general case: if $\mathbb{F}$ is a Galois field of characteristic p, if t is transcendental over $\mathbb{F}$ of degree prime to p and if $\mathbb{F}(t)$ is a field over the Steenrod algebra, then it is $\mathcal{P}^*$-inseparably closed.

PROOF: A Galois field $\mathbb{F}$ is perfect. Therefore we can apply the above Theorem 2.3.3 inductively starting with $\mathbb{K}^* = \mathbb{F}$. •

We return to the question we asked at the beginning of this section, namely, is there a "purely inseparable extension lemma". Let's take a *-purely inseparable finite field extension in the category of graded fields

$$\mathbb{L}^* = \mathbb{K}^*(l_1, \ldots, l_m)/\mathbb{K}^*,$$

where $\mathbb{K}^*$ enjoys a $\mathcal{P}^*$-action. Let's assume that all elements $l_i \in \mathbb{L}^*$ are roots of $\mathcal{P}^*$-purely inseparable polynomials. Take an $l \in \mathbb{L}^*$ whose minimal polynomial has degree p^e over $\mathbb{K}^*$. Then we can define the Steenrod algebra action on the additional element $l \in \mathbb{L}^*$ just as we did at the beginning of this section, where we extended the Steenrod algebra action successively to the fields $\mathbb{K}^*{}_i$. However, this construction does not necessarily lead to values in $\mathbb{L}^*$, i.e., if $l \in \mathbb{L}^*$ then we just don't know whether

$$\mathcal{P}^j_{\mathbb{L}^*}(l)$$

is in $\mathbb{L}^*$ or not, because we don't know whether $\mathbb{L}^*$ also contains the splitting field of

$$X^{p^e(\deg(l)+j(q-1))} - \left(\mathcal{P}^j_{\mathbb{L}^*}(l)\right)^{p^e} \in \mathbb{K}^*[X].$$

Finally, note that, if we have an algebraic field extension

$$\mathbb{K}^* \hookrightarrow \mathbb{L}^*$$

of fields over $\mathcal{P}^*$, then, analogous to the classical situation, we can form the $\mathcal{P}^*$-inseparable closure of $\mathbb{K}^*$ inside $\mathbb{L}^*$, $\mathbb{K}^*_{\mathcal{P}^*-\text{insep},\ \mathbb{L}^*/\mathbb{K}^*}$ such that

$$\mathbb{K}^* \underset{\mathcal{P}^*-\text{purely insep.}}{\hookrightarrow} \mathbb{K}^*_{\mathcal{P}^*-\text{insep},\ \mathbb{L}^*/\mathbb{K}^*} \underset{\mathcal{P}^*-\text{sep.}}{\hookrightarrow} \mathbb{L}^*.$$

This is indeed again a field closed under the action of the Steenrod algebra: Let $l \in \mathbb{L}^*$ be $\mathcal{P}^*$-purely inseparable over $\mathbb{K}^*$. Then there exists an e such that

$$l^{p^e} \in \mathbb{K}^*.$$

Hence for every $i \in \mathbb{N}_0$

$$\mathcal{P}^i(l)^{p^e} = \mathcal{P}^{ip^e}\left(l^{p^e}\right) \in \mathbb{K}^*,$$

i.e., any Steenrod power of our l is also $\mathcal{P}^*$-purely inseparable over $\mathbb{K}^*$

$$\mathcal{P}^i(l) \in \mathbb{K}^*_{\mathcal{P}^*-\text{insep},\ \mathbb{L}^*/\mathbb{K}^*}.$$

On the other hand, again, as in the classical case, we can take the $\mathcal{P}^*$-separable closure of $\mathbb{K}^*$ inside $\mathbb{L}^*$, $\mathbb{K}^*_{\mathcal{P}^*-\text{sep},\ \mathbb{L}^*/\mathbb{K}^*}$ such that

$$\mathbb{K}^* \underset{\mathcal{P}^*-\text{sep.}}{\hookrightarrow} \mathbb{K}^*_{\mathcal{P}^*-\text{sep},\ \mathbb{L}^*/\mathbb{K}^*} \underset{\mathcal{P}^*-\text{purely insep.}}{\hookrightarrow} \mathbb{L}^*.$$

As in the first case it is easily seen that the intermediate field is again closed under the action of the Steenrod algebra. If $l \in \mathbb{L}^*$ is $\mathcal{P}^*$-separable over $\mathbb{K}^*$, then

Proposition 2.2.2 provides us with a recursion formula for its Steenrod powers, which in turn leads to

$$\mathcal{P}^i(l) \in \mathbb{K}^* \left(l,\ \mathcal{P}^1(l), \dots,\ \mathcal{P}^{i-1}(l)\right)$$

and hence inductively

$$\mathcal{P}^i(l) \in \mathbb{K}^*(l) \in \mathbb{K}^*_{\mathcal{P}^*\text{-sep},\ \mathbb{L}^*/\mathbb{K}^*}$$

for all $i \in \mathbb{N}_0$.

2.4 The Vector Space of Derivations

Let $\mathbb{K}^*$ be a field over the Steenrod algebra. In Section 1.1 we defined the module of derivations $\Delta(\mathrm{H}^*)$ of an unstable algebra, H^*, over the Steenrod algebra. It is generated by $\mathcal{P}^{\Delta_0}$ and the Steenrod derivations $\mathcal{P}^{\Delta_i}$ for $i \geq 1$. The fact that these elements are derivations does not depend on unstability, but, in case of the $\mathcal{P}^{\Delta_i}$'s (where $i \geq 1$), is a consequence of the Adem-Wu relations and the Cartan formulae, while for $\mathcal{P}^{\Delta_0}$ it is obvious. This means the elements

$$\mathcal{P}^{\Delta_0},\ \mathcal{P}^{\Delta_1},\ \mathcal{P}^{\Delta_2},\ \dots$$

generate a **vector space of derivations of the field** $\mathbb{K}^*$. We use the same notation as for algebras, i.e., $\Delta(\mathbb{K}^*)$. By the remark on page 12 this vector space is closed under the Lie bracket. Moreover, for all $i \geq 1$, we have

$$\underbrace{\mathcal{P}^{\Delta_i} \cdots \mathcal{P}^{\Delta_i}}_{\leftarrow p\,-\,\text{times}\rightarrow} = 0,$$

as a consequence of the Adem-Wu relations. For $i = 0$, and for every $k \in \mathbb{K}^*$,

$$\underbrace{\mathcal{P}^{\Delta_0} \cdots \mathcal{P}^{\Delta_0}}_{\leftarrow p\,-\,\text{times}\rightarrow}(k) = \deg(k)^p k = \deg(k)k,$$

because the degree is an integer and we are in characteristic p. This means that $\Delta(\mathbb{K}^*)$ is also closed under taking p-th powers, i.e., $\Delta(\mathbb{K}^*)$ is a **restricted Lie algebra**, see page 186 of [17] for a definition.

THEOREM 2.4.1: *Let* $\mathbb{K}^* \hookrightarrow \mathbb{L}^*$ *be a field extension over the Steenrod algebra. If*

$$\dim_{\mathbb{K}^*}(\Delta(\mathbb{K}^*)) = \dim_{\mathbb{L}^*}(\Delta(\mathbb{L}^*)) = m < \infty$$

then either $\mathbb{K}^* = \mathbb{L}^*$ *or* $\mathbb{L}^*$ *is not* $\mathcal{P}^*$*-inseparably closed.*

PROOF: If $\mathbb{K}^* = \mathbb{L}^*$ nothing is to be shown. So assume that $\mathbb{K}^* \hookrightarrow \mathbb{L}^* = \mathbb{K}^*(\mathcal{S})$ is a non-trivial field extension, where $\mathcal{S}$ is a minimal generating set. Denote by $\mathbb{K}^*{}_c$, resp. $\mathbb{L}^*{}_c$, the subfield of $\Delta(\mathbb{K}^*)$-constants, resp. $\Delta(\mathbb{L}^*)$-constants. Then by Theorem 19 on page 186 of [17], the field extensions

$$\mathbb{K}^*{}_c \hookrightarrow \mathbb{K}^*,\ \text{and},\ \mathbb{L}^*{}_c \hookrightarrow \mathbb{L}^*$$

both are purely inseparable of degree p^m. Therefore

$$\mathbb{K}^*{}_c(\mathcal{S}) \hookrightarrow \mathbb{K}^*(\mathcal{S}) = \mathbb{L}^*$$

is purely inseparable of degree p^m. Since

$$\mathbb{L}^*{}_c \hookrightarrow \mathbb{K}^*{}_c(\mathcal{S}) \hookrightarrow \mathbb{L}^*$$

the two smaller fields must be equal. This, in turn, means that any element $s \in \mathcal{S}$ is a $\Delta(\mathbb{L}^*)$-constant and, since $\mathcal{S}$ was minimal, has no p-th root in $\mathbb{L}^*$. Therefore $\mathbb{L}^*$ is not $\mathcal{P}^*$-inseparably closed. •

We append a corollary about field extensions of rational function fields, which will be of use later. Note that the field of rational functions in n variables of degree one

$$\mathbb{F}(x_1, \ldots, x_n)$$

is the field of fractions of the polynomial ring $\mathbb{F}[x_1, \ldots, x_n]$. The Steenrod algebra acts on it by setting[12]

$$\boldsymbol{P}(\xi)\left(\frac{f}{h}\right) := \frac{\boldsymbol{P}(\xi)(f)}{\boldsymbol{P}(\xi)(h)},$$

and the Steenrod derivations satisfy the quotient rule of the calculus

$$\mathcal{P}^{\Delta_i}\left(\frac{f}{h}\right) = \frac{h\mathcal{P}^{\Delta_i}(f) - f\mathcal{P}^{\Delta_i}(h)}{h^2}.$$

COROLLARY 2.4.2: *Let $\mathbb{F}(x_1, \ldots, x_n) \hookrightarrow \mathbb{L}^*$ be a field extension of fields over the Steenrod algebra, and let*

$$\dim_{\mathbb{L}^*}(\Delta(\mathbb{L}^*)) = n.$$

$\mathbb{L}^*$ is $\mathcal{P}^*$-inseparably closed if and only if $\mathbb{F}(x_1, \ldots, x_n) = \mathbb{L}^*$.

PROOF: If $\mathbb{L}^*$ is not $\mathcal{P}^*$-inseparably closed, it can't be equal to $\mathbb{F}(x_1, \ldots, x_n)$, because the smaller field is $\mathcal{P}^*$-inseparably closed, by Corollary 2.3.4.

For the converse, assume that $\mathbb{L}^*$ is $\mathcal{P}^*$-inseparably closed. By Theorem 1.2.3, the module of derivations, $\Delta(\mathbb{F}[x_1, \ldots, x_n])$, is a free module over the polynomial ring generated by n elements, Hence the vector space of derivations, $\Delta(\mathbb{F}(x_1, \ldots, x_n))$, has also dimension n. So the preceding theorem applies. •

[12] We will come to this in more detail in the following chapter.

Chapter 3
The Integral Closure Theorem and the Unstable Part

In this chapter we are going to generalize the Integral Closure Theorem[1] of Clarence W. Wilkerson, [44], to arbitrary Galois fields.

Throughout the whole chapter H^* is an unstable algebra over the Steenrod algebra.

3.1 Rings of Fractions and Their Unstable Part

Let $S \subseteq H^*$ be a multiplicatively closed subset, and form the ring of fractions

$$S^{-1}H^* := \left\{ \frac{h}{s} \mid h \in H^*,\ s \in S \right\}.$$

There is an extension of the action of the Steenrod algebra to $S^{-1}H^*$ given by requiring

$$\mathbf{P}(\xi)\left(\frac{h}{s}\right)\mathbf{P}(\xi)(s) = \mathbf{P}(\xi)(h)$$

for any $\frac{h}{s} \in S^{-1}H^*$, compare[2] [43] Proposition 2.1. By construction, this action satisfies the Cartan formulae, but not the unstability condition.

It will be useful to recall from [14], Definition 2.2, the notion of the unstable part:

Definition: *Let M^* be a graded module over the Steenrod algebra. For a multi index $I = (i_1, \ldots, i_s)$ denote by $\mathcal{P}^I$ the product $\mathcal{P}^{i_1} \cdots \mathcal{P}^{i_s}$. Then the*

---

[1] Wilkerson calls this theorem the "Generalized Serre Lemma", because it extends a result from J.-P. Serre, [33], in some way.

[2] Or do it yourself: the proof works very similar to the one on page 37 where we extended a $\mathcal{P}^*$-action from a smaller field to a bigger one.

unstable part of M^* *in degree* k *is the* $\mathbb{F}$*-vector subspace defined by*

$$\left(\mathcal{U}n(\mathrm{M}^*)\right)_{(k)} := \left\{m \in \mathrm{M}^*_{(k)} \mid \mathcal{P}^r \mathcal{P}^I(m) = 0,\, r > k + \deg(\mathcal{P}^I),\, \forall \text{ multi indices } I\right\}.$$

The unstable part is a graded submodule over $\mathcal{P}^*$, which satisfies the unstability condition (compare again [14]). Moreover, the unstable part of an *algebra* over the Steenrod algebra is a sub*algebra* over the Steenrod algebra.

We need the following result, compare Proposition 2.2 in [44].

LEMMA 3.1.1: *Let* $\mathbb{L}^*/\mathbb{K}^*$ *be a* $\mathcal{P}^*$*-algebraic extension of graded fields over the Steenrod algebra. Then the ring extension*

$$\mathcal{U}n(\mathbb{K}^*) \hookrightarrow \mathcal{U}n(\mathbb{L}^*)$$

is integral.

PROOF: As explained at the end of Section 2.3 we form[3] the $\mathcal{P}^*$-separable closure $\mathbb{K}^*{}_{\mathcal{P}^*-\text{sep},\mathbb{L}^*/\mathbb{K}^*}$ of $\mathbb{K}^*$ in $\mathbb{L}^*$. We have then two field extensions

$$\mathbb{K}^* \underset{\mathcal{P}^*-\text{sep}}{\hookrightarrow} \mathbb{K}^*{}_{\mathcal{P}^*-\text{sep},\, \mathbb{L}^*/\mathbb{K}^*} \underset{\mathcal{P}^*-\text{purely insep}}{\hookrightarrow} \mathbb{L}^*$$

and it is enough to show that both of the induced extensions of the unstable parts

$$\mathcal{U}n(\mathbb{K}^*) \hookrightarrow \mathcal{U}n(\mathbb{K}^*{}_{\mathcal{P}^*-\text{sep},\, \mathbb{L}^*/\mathbb{K}^*}) \hookrightarrow \mathcal{U}n(\mathbb{L}^*)$$

are integral. So, take an element

$$l \in \mathcal{U}n(\mathbb{L}^*) \subset \mathbb{L}^*.$$

Then l is $\mathcal{P}^*$-purely inseparable over $\mathbb{K}^*{}_{\mathcal{P}^*-\text{sep},\, \mathbb{L}^*/\mathbb{K}^*}$. Hence there exists an $s \in \mathbb{N}$ such that

$$l^{q^s} \in \mathbb{K}^*{}_{\mathcal{P}^*-\text{sep},\, \mathbb{L}^*/\mathbb{K}^*}.$$

However, this element is again unstable, since it is just a q-th power of an unstable element, i.e.,

$$l^{q^s} \in \mathcal{U}n(\mathbb{K}^*{}_{\mathcal{P}^*-\text{sep},\, \mathbb{L}^*/\mathbb{K}^*})$$

and l is integral over $\mathcal{U}n(\mathbb{K}^*{}_{\mathcal{P}^*-\text{sep},\, \mathbb{L}^*/\mathbb{K}^*})$. Next take an $l \in \mathcal{U}n(\mathbb{K}^*{}_{\mathcal{P}^*-\text{sep},\, \mathbb{L}^*/\mathbb{K}^*})$. It is, by construction, separable over $\mathbb{K}^*$, and hence its minimal polynomial

$$\text{minpol}_{l \in \mathbb{K}^*{}_{\mathcal{P}^*-\text{sep}}/\mathbb{K}^*}(X) = \kappa_0 + \kappa_1 X + \cdots + X^m \in \mathbb{K}^*[X]$$

is separable. Lets take its splitting field $\Delta^*/\mathbb{K}^*$, i.e.,

$$\kappa_0 + \kappa_1 X + \cdots + X^m = \prod_{i=1}^m (X - r_i) \in \Delta^*[X],$$

[3] We could have also taken the $\mathcal{P}^*$-inseparable closure of $\mathbb{K}^*$ in $\mathbb{L}^*$. The proof would work just equally fine.

where without loss of generality $r_1 = l$. Any automorphism of Δ^* fixing $\mathbb{K}^*$ commutes with the action of the Steenrod algebra, because the extension of the Steenrod algebra action from $\mathbb{K}^*$ to Δ^* is unique by Proposition 2.2.2, compare Proposition 1.1 (c) in [44]. Hence, if one root is an unstable element, then all roots are unstable. Therefore the coefficients $\kappa_0, \ldots, \kappa_{m-1}$ are unstable elements, because they are just elementary symmetric functions in the roots, i.e., the minimal polynomial has unstable coefficients

$$\mathrm{minpol}_{l \in \mathbb{K}^*_{\mathcal{P}^*\text{-sep}}/\mathbb{K}^*}(X) = \kappa_0 + \kappa_1 X + \cdots + X^m \in \mathcal{U}n(\mathbb{K}^*)[X],$$

which in turn means that l is integral over $\mathcal{U}n(\mathbb{K}^*)$. •

For rings of fractions the unstable part can be explicitly calculated:

Lemma 3.1.2: *Let* H^* *be an unstable algebra over the Steenrod algebra. Let* $S^{-1}\mathrm{H}^*$ *be the ring of fractions for a multiplicatively closed subset* $S \subseteq \mathrm{H}^*$. *Then the unstable part of* $S^{-1}\mathrm{H}^*$ *is*

$$\mathcal{U}n(S^{-1}\mathrm{H}^*) = \left\{\frac{h}{s}, \text{ such that } \boldsymbol{P}(\xi)(s) \mid \boldsymbol{P}(\xi)(h)\right\}.$$

Proof: Let $\frac{h}{s} \in S^{-1}\mathrm{H}^*$. Then the condition

$$\boldsymbol{P}(\xi)(s) \mid \boldsymbol{P}(\xi)(h)$$

is equivalent to the statement that

$$\boldsymbol{P}(\xi)\left(\frac{h}{s}\right) = \frac{\boldsymbol{P}(\xi)(h)}{\boldsymbol{P}(\xi)(s)}$$

is a *polynomial* in ξ of degree $\deg(h) - \deg(s)$ with highest coefficient $\left(\frac{h}{s}\right)^q$. This is just another way of saying that $\frac{h}{s}$ is unstable. •

An integrally closed unique factorization domain is equal to the unstable part of its field of fractions (compare Proposition 3.3 in [44])[4], as the following lemma shows.

Lemma 3.1.3: *If* H^* *is an unstable, integrally closed, unique factorization domain, then*

$$\mathrm{H}^* = \mathcal{U}n(FF(\mathrm{H}^*)).$$

[4] A more general investigation of the relation of an unstable algebra H^* and the unstable part of any of its localizations can be found in [27].

PROOF: Let

$$\frac{h}{s} \in \mathcal{U}n(FF(\mathrm{H}^*)).$$

By definition of the unstable part this means that

$$\boldsymbol{P}(\xi)(s) \mid \boldsymbol{P}(\xi)(h) \in FF(\mathrm{H}^*)[[\xi]],$$

and we have to show that s divides h in H^*. Since we are working in a unique factorization domain there is no loss of generality assuming that $s \in \mathrm{H}^*$ is a prime element. Define the s-content of a polynomial $f(\xi) = h_0 + h_1\xi + \cdots + h_m\xi^m \in \mathrm{H}^*[\xi]$ by

$$c_s(f) := s^i,$$

where

$$i := \max\{ j \in \mathbb{N}_0 \text{ such that } s^j \mid h_0, \ldots, s^j \mid h_m\}.$$

We find that the s-content of

$$\boldsymbol{P}(\xi)(s) = s + \mathcal{P}^1(s)\xi + \cdots + s^q\xi^{\deg(s)}$$

is s^0 or s^1. Assume that s does not divide h. Then the s-content of

$$\boldsymbol{P}(\xi)\left(\frac{h}{s}\right) = \frac{h}{s} + \cdots + \left(\frac{h}{s}\right)^q \xi^{\deg(h)-\deg(s)}$$

is s^{-k} for some large $k \geq q$. Since

$$\boldsymbol{P}(\xi)(s)\boldsymbol{P}(\xi)\left(\frac{h}{s}\right) = \boldsymbol{P}(\xi)(h)$$

we get that the s-content of $\boldsymbol{P}(\xi)(h)$ is

$$s^{-k+0} \quad \text{or} \quad s^{-k+1},$$

but this is a contradiction, because $-k + 0 < -k + 1 < 0$. So s divides h. •

3.2 The Integral Closure Theorem

We come next to the generalization of C. W. Wilkerson's Integral Closure Theorem. The following lemma is the main ingredient.

LEMMA 3.2.1: *Consider the polynomial algebra $\mathbb{F}[x_1, \ldots, x_n]$ over $\mathbb{F}$, where the generators x_i all have degree one. Let H^* be an integrally closed integral domain and*

$$\varphi : \mathbb{F}[x_1, \ldots, x_n] \hookrightarrow \mathrm{H}^* = \overline{\mathrm{H}^*}$$

an integral extension of unstable algebras over the Steenrod algebra. Then any linear form $l \in \mathbb{F}[x_1, \ldots, x_n]$ generates a prime ideal in H^*.

Proof: Without loss of generality we can assume that $l = x_1$. Let $h_1,\ h_2 \in$ H^* such that

$$h_1 h_2 = h x_1 \in (x_1)^e \subset \mathrm{H}^*,$$

where $(x_1)^e$ denotes the extension to H^* of the ideal $(x_1) \subset \mathbb{F}[x_1, \dots, x_n]$. We apply the giant Steenrod operation and get

$$\boldsymbol{P}(\xi)(h_1)\boldsymbol{P}(\xi)(h_2) = \boldsymbol{P}(\xi)(h)\boldsymbol{P}(\xi)(x_1) \in \mathrm{H}^*[\xi] \subset FF(\mathrm{H}^*)[\xi].$$

As an element in $FF(\mathrm{H}^*)[\xi]$

$$\boldsymbol{P}(\xi)(x_1) = x_1 + x_1^q \xi$$

is prime, because it is linear in ξ. Hence in $FF(\mathrm{H}^*)[\xi]$ it divides one of the factors on the left hand side, say

$$\boldsymbol{P}(\xi)(x_1) = x_1 + x_1^q \xi \mid \boldsymbol{P}(\xi)(h_1).$$

This means by definition that

$$\frac{h_1}{x_1} \in \mathcal{U}n(FF(\mathrm{H}^*)).$$

By Lemma 3.1.1 the unstable part $\mathcal{U}n(FF(\mathrm{H}^*))$ is integral over the unstable part $\mathcal{U}n\left(\mathbb{F}(x_1, \dots, x_n)\right)$. Lemma 3.1.3 tells us that

$$\mathcal{U}n(\mathbb{F}(x_1, \dots, x_n)) = \mathbb{F}[x_1, \dots, x_n].$$

Combining the two statements gives us that $\mathcal{U}n\,(FF(\mathrm{H}^*))$ is integral over $\mathbb{F}[x_1, \dots, x_n]$, i.e., $\frac{h_1}{x_1}$ is integral over $\mathbb{F}[x_1, \dots, x_n]$ and a fortiori over H^*. Since we assumed H^* to be integrally closed this means that $\frac{h_1}{x_1} \in \mathrm{H}^*$ or

$$h_1 \in (x_1)^e \subset \mathrm{H}^*,$$

and therefore $(x_1)^e \subset \mathrm{H}^*$ is a prime ideal. •

The proof of the following theorem is due to Larry Smith, [38], and simplifies considerably my original proof.

Theorem 3.2.2 (Integral Closure Theorem, General Version): *Consider the polynomial algebra $\mathbb{F}[x_1, \dots, x_n]$ over $\mathbb{F}$, where the generators x_i all have degree one. Let H^* be an integral domain and let*

$$\varphi : \mathbb{F}[x_1, \dots, x_n] \hookrightarrow \mathrm{H}^*$$

be an integral extension of unstable algebras over the Steenrod algebra. Then φ is an isomorphism of unstable algebras over the Steenrod algebra.

PROOF: Since

$$\mathbb{F}[x_1, \ldots, x_n] \underset{\varphi}{\hookrightarrow} \mathrm{H}^* \hookrightarrow \overline{\mathrm{H}^*}$$

is again an integral extension of unstable integral domains over $\mathcal{P}^*$ we assume without loss of generality that H^* is integrally closed.

We proceed by induction on the Krull dimension n. If $n = 0$ then $\mathbb{F} = \mathrm{H}^*$ since H^* is connected[5]. Let $n > 0$. We know from Lemma 3.2.1 that

$$(x_1)^e \subset \mathrm{H}^*$$

is a prime ideal. Since

$$(x_1)^e \cap \mathbb{F}[x_1, \ldots, x_n] = (x_1)$$

it follows that $(x_1)^e$ lies over (x_1). We therefore obtain a diagram

$$\begin{array}{ccccccccc} 0 \longrightarrow & \mathbb{F}[x_1, \ldots, x_n] & \stackrel{\mu}{\longrightarrow} & \mathbb{F}[x_1, \ldots, x_n] & \longrightarrow & \mathbb{F}[x_2, \ldots, x_n] & \longrightarrow 0 \\ & \downarrow \varphi & & \downarrow \varphi & & \downarrow \psi & \\ 0 \longrightarrow & \mathrm{H}^* & \stackrel{\mu}{\longrightarrow} & \mathrm{H}^* & \longrightarrow & \mathrm{H}^*/(x_1)^e & \longrightarrow 0 \end{array}$$

where the maps marked μ denote multiplication by x_1 and the map ψ is induced by the inclusion φ. By Lemma 3.2.1 $\mathrm{H}^*/(x_1)^e$ is an integral domain. So ψ is an isomorphism by the induction hypothesis. Hence by the serpent lemma

$$\mu : \operatorname{coker}(\varphi) \longrightarrow \operatorname{coker}(\varphi)$$

is an isomorphism. Since μ has degree +1 and $\operatorname{coker}(\varphi)$ is non-negatively graded it follows that $\operatorname{coker}(\varphi) = 0$ and we are done. •

REMARK: Note that this theorem does not remain valid if we drop the assumption on H^* to be an integral domain. Consider e.g.

$$\mathbb{F}[x] \hookrightarrow \mathbb{F}[x, y]/(y^2)$$

where x, y are linear generators.

Note carefully that the preceding theorem depends heavily on unstability. This is illustrated by the next example.

EXAMPLE 1: Let $\mathbb{F} = \mathbb{F}_3$ be the field with three elements and let $\mathbb{F}[x, y]$ be a polynomial ring with two linear generators. Embed $\mathbb{F}[x, y]$ in its field of fractions $\mathbb{F}(x, y)$ and extend this field to

$$\mathbb{K}^* := \mathbb{F}(x, y, t),$$

where $t^2 = xy$. We receive a $\mathcal{P}^*$-separable finite field extension

$$\begin{array}{ccl} \mathbb{F}[x, y] & & \\ \downarrow & & \\ \mathbb{F}(x, y) & \underset{\mathcal{P}^*\text{-sep}}{\hookrightarrow} & \mathbb{K}^* := \mathbb{F}(x, y, t), \quad t^2 = xy. \end{array}$$

[5] Recall from the introduction that a graded commutative $\mathbb{F}$-algebra is called connected, if the degree zero part is precisely the ground field $\mathbb{F}$.

Proposition 2.2.2 tells us how to extend the $\mathcal{P}^*$-action naturally given on $\mathbb{F}(x, y)$ uniquely to $\mathbb{K}^*$, namely by solving the equations

$$2\mathcal{P}^i(t)t + \sum_{r+s=i;\ r,s<i} \mathcal{P}^r(t)\mathcal{P}^s(t) = \mathcal{P}^i(t^2) = \mathcal{P}^i(xy)$$

inductively for $\mathcal{P}^i$, e.g.,

$$\mathcal{P}^1(t) = -\frac{1}{t}\mathcal{P}^1(xy) = -\frac{1}{t}(x^3y + xy^3) = -\frac{1}{t}t^2(x^2 + y^2) = -t(x^2 + y^2)$$

and

$$\begin{aligned}\mathcal{P}^2(t) &= -\frac{1}{t}\left(\mathcal{P}^2(xy) - \mathcal{P}^1(t)^2\right) = -\frac{1}{t}\left(x^3y^3 - t^2(x^2 + y^2)^2\right)\\ &= \frac{1}{t}\left(t^6 + t^2x^4 + t^2y^4\right) = t\left(t^4 + x^4 + y^4\right)\\ &= t\left(x^2 - y^2\right)^2\end{aligned}$$

So, the $\mathcal{P}^*$-action on t is no longer unstable, for otherwise we would have

$$txy = t^3 = \mathcal{P}^1(t) = -t(x^2 + y^2),$$

and hence

$$xy = -(x^2 + y^2),$$

which contradicts the fact that $\mathbb{F}(x, y)$ is a purely transcendental field extension. Also, we obviously have

$$0 \neq \mathcal{P}^2(t) = t\left(x^2 - y^2\right)^2.$$

So, let's take the unstable part of $\mathbb{K}^*$. Then we have a diagram in the category of fields, resp. unstable algebras over $\mathcal{P}^*$ of the following form

$$\begin{array}{ccc}\mathbb{F}[x, y] & \overset{\varphi}{\hookrightarrow} & \mathcal{U}n(\mathbb{K}^*)\\ \downarrow & & \downarrow\\ \mathbb{F}(x, y) & \underset{\mathcal{P}^*\text{-sep, finite}}{\hookrightarrow} & \mathbb{K}^*.\end{array}$$

Since $\mathcal{U}n(\mathbb{F}(x, y)) = \mathbb{F}[x, y]$ by Lemma 3.1.3, we can apply Lemma 3.1.1 to conclude that φ is an integral ring extension. So by Theorem 3.2.2 the map φ is an isomorphism and

$$\mathcal{U}n(\mathbb{K}^*) = \mathbb{F}[x, y].$$

However, let's ignore the Steenrod algebra for a moment and take, instead of the unstable part of $\mathbb{K}^*$, the integral closure of $\mathbb{F}[x, y]$ in $\mathbb{K}^*$: this gives

$$\begin{array}{ccc}\mathbb{F}[x, y] & \overset{\psi}{\hookrightarrow} & \overline{\mathbb{F}[x, y]}_{\mathbb{K}^*}\\ \downarrow & & \downarrow\\ \mathbb{F}(x, y) & \underset{\mathcal{P}^*\text{-sep, finite}}{\hookrightarrow} & \mathbb{K}^*.\end{array}$$

Then, by construction, the map ψ is an integral ring extension, but it is far from being an isomorphism, because

$$\overline{\mathbb{F}[x,\ y]}_{\mathbb{K}^*} = \mathbb{F}[x,\ y,\ t]/(t^2 - xy)$$

as one should easily see. The reason is, that this new ring is *not unstable* over the Steenrod algebra: recall from above that we found

$$\mathcal{P}^1(t) \neq t^3$$

and, e.g.,

$$\mathcal{P}^2(t) \neq 0.$$

We close this chapter with an obvious corollary for further reference.

Recall that a ring extension $R \hookrightarrow S$ is called **algebraic**, if every element $s \in S$ is algebraic over R, i.e., satisfies an algebraic relation with coefficients in S. The **algebraic closure** of R is maximal algebraic overring over R, and any other algebraic overring over R is contained in the algebraic closure.

COROLLARY 3.2.3: *A polynomial ring* $\mathbb{F}[x_1, \ldots, x_n]$ *with linear generators* $x_1, \ldots, x_n$ *is algebraically closed, in the category of unstable integral domains. The algebraic closure of the Dickson algebra* $\mathcal{D}^*(n)$ *is* $\mathbb{F}[x_1, \ldots, x_n]$.

PROOF: Let

$$\varphi : \mathbb{F}[x_1, \ldots, x_n] \hookrightarrow \mathrm{H}^*$$

be an algebraic extension of unstable integral domains. Then the induced map between the respective field of fractions

$$\Phi : \mathbb{F}(x_1, \ldots, x_n) \hookrightarrow FF(\mathrm{H}^*)$$

is also algebraic. Therefore, by Lemma 3.1.1 and Lemma 3.1.3 the map

$$\varphi : \mathcal{U}n(\mathbb{F}(x_1, \ldots, x_n)) = \mathbb{F}[x_1, \ldots, x_n] \hookrightarrow \mathrm{H}^* \hookrightarrow \mathcal{U}n(FF(\mathrm{H}^*))$$

is integral. Therefore, by Theorem 3.2.2, φ is an isomorphism, and $\mathbb{F}[x_1, \ldots, x_n]$ is algebraically closed.

For the second assertion, note that

$$\mathcal{D}^*(n) \hookrightarrow \mathbb{F}[x_1, \ldots, x_n]$$

is an algebraic extension. Therefore the algebraic closure of the Dickson algebra contains $\mathbb{F}[x_1, \ldots, x_n]$, which in turn is algebraically closed. •

Chapter 4
The Inseparable Closure

After having settled the awful technical stuff we are about to arrive at our first success: the discovery of a new exotic animal, the $\mathcal{P}^*$-inseparable closure of an algebra H^* over the Steenrod algebra $\mathcal{P}^*$. Until now this has been known only for the prime field $\mathbb{F}_p$, and even there, only for Noetherian unstable integral domains. However, as we show, the $\mathcal{P}^*$-inseparable closure can be constructed more generally. We just have to be more careful and patient.

Throughout the chapter H^* is just a graded connected commutative $\mathbb{F}$-algebra. Any further assumptions on H^* will be explicitly stated when needed.

4.1 The Introduction of our Exotic Animal

Let's start with a definition.

DEFINITION: *Call an algebra* H^* *over the Steenrod algebra* $\mathcal{P}^*$**-inseparably closed**, *if whenever* $h \in \mathrm{H}^*$ *and*

$$\mathcal{P}^{\Delta_i}(h) = 0 \quad \forall\, i \geq 0,$$

then there exists an element $h' \in \mathrm{H}^*$ *such that*

$$(h')^p = h.$$

In other words, if all our derivations $\mathcal{P}^{\Delta_i}$ vanish on an element in H^* that element must be a p-th power[1]. Before elaborating a bit more let's append the obvious definition of a $\mathcal{P}^*$-inseparable closure.

DEFINITION: *The* $\mathcal{P}^*$**-inseparable closure** *of* H^* *is a* $\mathcal{P}^*$*-inseparably closed algebra* $\sqrt[\mathcal{P}^*]{\mathrm{H}^*}$ *containing* H^* *such that the following universal property*[2]

[1] Note that we need to detect p-th powers, not just q-th powers.

[2] I agree this is just some abstract nonsense, but wait a minute and we will see that this animal really exists.

holds: Whenever we have a $\mathcal{P}^*$-inseparably closed algebra H'^* containing H^* there exists an embedding $\varphi : \sqrt[\mathcal{P}^*]{\mathrm{H}^*} \hookrightarrow \mathrm{H}'^*$.

S.-P. Lam defined the $\mathcal{P}^*$-inseparable closure in the category of unstable Noetherian integral domains over the Steenrod algebra, [19] §3. Moreover, he showed that his definition is equivalent to condition 1.2.2 occurring in J. F. Adam's and C. W. Wilkerson's proof that certain unstable integral domains over the Steenrod algebra are rings of invariants, see again [19] Proposition 3.2 and compare [1]. However, the above way to define $\sqrt[\mathcal{P}^*]{\mathrm{H}^*}$ gives the definition wider scope: Note that we not only drop the assumption on H^* being Noetherian and containing no zero divisors, but also we don't need unstability.

It is clear that, if $\sqrt[\mathcal{P}^*]{\mathrm{H}^*}$ exists, then it is unique. However, the hard part begins now: we have to show that $\sqrt[\mathcal{P}^*]{\mathrm{H}^*}$ exists. We do that by giving a *construction method*, which mimics the construction method of the $\mathcal{P}^*$-inseparable closure of a field given in Section 2.3. Note that our method does *not* work for algebras with nilpotent elements, i.e., we can't prove the existence of $\sqrt[\mathcal{P}^*]{\mathrm{H}^*}$ for nonreduced algebras.

Denote by $C^* \subseteq \mathrm{H}^*$ the subalgebra consisting of the $\mathcal{P}^{\Delta_i}$ constants for all $i \geq 0$, i.e.,

$$C^* = C(\mathrm{H}^*) = \{h \in \mathrm{H}^* \mid \mathcal{P}^{\Delta_i}(h) = 0 \ \forall\ i \geq 0\}.$$

Note that the nonzero elements of C^* all have degree divisible by p, because

$$\mathcal{P}^{\Delta_0}(h) = \deg(h)h.$$

LEMMA 4.1.1: *If* H^* *is Noetherian then* C^* *is a finitely generated algebra over* $\mathbb{F}$.

PROOF: We have that the extension

$$C^* \hookrightarrow \mathrm{H}^*$$

is finite, since for any $h \in \mathrm{H}^*$ its p-th power is in C^*, $h^p \in C^*$ and H^* is Noetherian. Therefore C^* is a Noetherian ring by a classic result due to Emmy Noether, see e.g. proof of Theorem 2.3.1 in [36]. Hence C^* is finitely generated by standard yoga for graded connected commutative $\mathbb{F}$-algebras, see e.g. Section 2.2 in [36]. •

The following lemma ensures that C^* inherits an action of the Steenrod algebra from H^*.

LEMMA 4.1.2: *C^* is closed under the action of the Steenrod algebra $\mathcal{P}^*$.*

PROOF: Let $h \in C^*$. We have to show that

$$\mathcal{P}^j(h) \in C^* \quad \forall\, j \geq 0,$$

i.e.,

$$\mathcal{P}^{\Delta_i}\mathcal{P}^j(h) = 0 \quad \forall\, i,\, j \geq 0.$$

First we show this for all $i \geq 1$ by induction on $j \geq 0$. Since $\mathcal{P}^0 = \mathrm{id}$ the induction starts by hypothesis. We have for every $i \geq 1$ and for every $j \geq 0$

$$\mathcal{P}^{\Delta_i}\left(\mathcal{P}^j(h)\right) = \mathcal{P}^j\left(\mathcal{P}^{\Delta_i}(h)\right) - \mathcal{P}^{\Delta_{i+1}}\left(\mathcal{P}^{j-q^i}(h)\right) = -\mathcal{P}^{\Delta_{i+1}}\left(\mathcal{P}^{j-q^i}(h)\right) = 0,$$

where we made use of the commutation rules given by Lemma 1.1.2, the assumption that $\mathcal{P}^{\Delta_i}(h) = 0$ for all $h \in C^*$ and all $i \in \mathbb{N}_0$, the convention that $\mathcal{P}^i = 0$ if $i \notin \mathbb{N}_0$ and the induction, since $j - q^i < j$. Finally, to show that

$$\mathcal{P}^{\Delta_0}\left(\mathcal{P}^j(h)\right) = 0 \quad \forall\, j \geq 0$$

we consider three cases:

$j \equiv 0 \text{ MOD } p,$
$j \equiv 1 \text{ MOD } p,$ and
$j \not\equiv 0 \text{ or } 1 \text{ MOD } p.$

Let $h \in C^*$. Then $\deg(h) \equiv 0 \text{ MOD } p$, because $\mathcal{P}^{\Delta_0}(h) = 0$. Hence for $j \equiv 0 \text{ MOD } p$ we have

$$\deg\left(\mathcal{P}^j(h)\right) = \deg(h) + j(q-1) \equiv 0 \text{ MOD } p.$$

On the other hand, if j is not divisible by p we prove by induction that $\mathcal{P}^j(h) = 0$. For $j = 1$ this is true by definition of C^*, because $\mathcal{P}^1 = \mathcal{P}^{\Delta_1}$. Let $j > 1$ and not divisible by p. Consider the following Adem-Wu relation:

$$j\mathcal{P}^j = \mathcal{P}^1\mathcal{P}^{j-1}.$$

If $j \equiv 1 \text{ MOD } p$ then $\mathcal{P}^{j-1}(h) \in C^*$ by what we have proved above. Hence

$$\mathcal{P}^j(h) = \frac{1}{j}\mathcal{P}^{\Delta_1}\mathcal{P}^{j-1}(h) = 0.$$

If $j \not\equiv 1 \text{ MOD } p$ then, by induction,

$$\mathcal{P}^{j-1}(h) = 0,$$

and hence, again,

$$\mathcal{P}^j(h) = \frac{1}{j}\mathcal{P}^{\Delta_1}\mathcal{P}^{j-1}(h) = 0,$$

as claimed. •

We want to construct a new algebra H_1^* with suitable properties, so that every element in C^* has a p-th root in H_1^*. Denote by

$$\Phi : \mathrm{H}^* \longrightarrow \mathrm{H}^*,\ h \longmapsto h^p$$

the Frobenius map. By construction we have integral extensions of algebras over the Steenrod algebra

$$(\mathrm{H}^*)^p := \Phi(\mathrm{H}^*) \hookrightarrow C^* \hookrightarrow \mathrm{H}^*.$$

Denote by $\mathcal{S}$ a set of generators for C^* as a module over $(\mathrm{H}^*)^p$. Define an algebra

$$\begin{aligned}\mathrm{H}_1^* &:= \mathrm{H}^*[\gamma_s \mid s \in \mathcal{S}]/\mathcal{R}ad\left(\gamma_s^p - s \mid s \in \mathcal{S}\right)\\ &= \left(\mathrm{H}^* \otimes_{\mathbb{F}} \mathbb{F}[\gamma_s \mid s \in \mathcal{S}]\right) / \mathcal{R}ad\left(\gamma_s^p - s \mid s \in \mathcal{S}\right).\end{aligned}$$

Note that this construction comes with a canonical map

$$\varphi_0 : \mathrm{H}^* \longrightarrow \mathrm{H}_1^*.$$

We start by proving the basic properties of H_1^*.

LEMMA 4.1.3 : *Let H^* be an algebra over the Steenrod algebra with $\mathcal{P}^{\Delta_i}$-constants $C^* \subseteq \mathrm{H}^*$. Then the algebra defined above H_1^* together with the canonical map φ_0, has the following properties.*

(1) *H_1^* is a reduced graded connected commutative Noetherian algebra over $\mathbb{F}$.*
(2) *If H^* is Noetherian, then so is H_1^*, and H_1^* is finitely generated as H^*-module.*
(3) *If $h \in \mathrm{H}_1^*$ then $h^p \in C^*$.*
(4) *If $c \in C^*$ then there exists an element $\gamma \in \mathrm{H}_1^*$ such that $\gamma^p = c$.*
(5) *If H^* is reduced then φ_0 is monic.*

PROOF: We take the things in order.

AD (1) : Give the new generators γ_s the obvious degree, namely

$$\deg(\gamma_s) := \frac{1}{p}\deg(s) \quad \forall\ s.$$

The condition

$$\mathcal{P}^{\Delta_0}(s) = \deg(s)s = 0 \quad \forall\ s$$

ensures that $\deg(s)$ is indeed divisible by p, so our new object is again graded over the non-negative integers. Connectivity and commutativity is free of charge. Finally, the algebra H_1^* is reduced by construction.

AD (2) : If H^* is Noetherian, then, by Lemma 4.1.1 C^* is finitely generated as an $\mathbb{F}$-algebra. Since

$$(\mathrm{H}^*)^p \hookrightarrow C^* \hookrightarrow \mathrm{H}^*$$

is a finite extension, C^* is also a finitely generated $(\mathrm{H}^*)^p$-module. Let $c_1, \ldots, c_t$ be a set of module generators. Then the algebra H_1^* is, as an H^*-module, generated by the elements γ_i^j for $i = 1, \ldots, t$ and $j = 1, \ldots, p-1$, because $\gamma_i^p = c_i \in \mathrm{H}^*$ by construction.

AD (3) : Let

$$h = \sum_{\text{finite}} h_{(i_1,\dots,i_{t_j})}\gamma_1^{i_1}\cdots\gamma_{t_j}^{i_{t_j}} \in \mathrm{H}_1^*$$

be an arbitrary element, where $h_{(i_1,\dots,i_{t_j})} \in \mathrm{H}^*$. Then

$$h^p = \sum_{\text{finite}} h^p_{(i_1,\dots,i_{t_j})}\gamma_1^{pi_1}\cdots\gamma_{t_j}^{pi_{t_j}} = \sum_{\text{finite}} h^p_{(i_1,\dots,i_{t_j})}c_1^{i_1}\cdots c_{t_j}^{i_{t_j}} \in C^*.$$

AD (4) : Every element $c \in C^*$ can be written in the form

$$c = h_1^p s_1 + \cdots + h_{t(c)}^p s_{t(c)}$$

for suitable $h_1,\dots, h_{t(c)} \in \mathrm{H}^*$, $s_1,\dots, s_{t(c)} \in \mathcal{S}$. Hence in H_1^* we have

$$c = h_1^p s_1 + \cdots + h_{t(c)}^p s_{t(c)} = h_1^p\gamma_{s_1}^p + \cdots + h_{t(c)}^p\gamma_{s_{t(c)}}^p = (h_1\gamma_{s_1} + \cdots + h_{t(c)}\gamma_{s_{t(c)}})^p,$$

i.e., the element $h_1\gamma_{s_1} + \cdots + h_{t(c)}\gamma_{s_{t(c)}} \in \mathrm{H}_1^*$ is the p-th root of c.

AD (5) : If $\varphi_0(h) = 0$ for some $h \in \mathrm{H}^*$, then

$$\varphi_0(h) \in \mathcal{Rad}(\gamma_s^p - s \mid s \in \mathcal{S}) \subseteq \mathrm{H}^*[\gamma_s \mid s \in \mathcal{S}].$$

Hence there exists an integer k such that

$$h^k = \varphi_0(h)^k \in (\gamma_s^p - s \mid s \in \mathcal{S}) \subseteq \mathrm{H}^*[\gamma_s \mid s \in \mathcal{S}],$$

i.e.,

$$h^k \in (\gamma_s^p - s \mid s \in \mathcal{S}) \cap \mathrm{H}^* = (0),$$

so h^k is nilpotent. So if H^* is reduced $h = 0$. •

Remark: If H^* is not reduced, then $\mathrm{H}^*/\mathcal{Nil}(\mathrm{H}^*) \hookrightarrow \mathrm{H}_1^* = \left(\mathrm{H}^*/\mathcal{Nil}(\mathrm{H}^*)\right)_1$.

We need an action of the Steenrod algebra on H_1^*. This follows from the next Lemma.

Lemma 4.1.4: *Suppose that $\mathrm{H}'^* \hookrightarrow \mathrm{H}''^*$ is an inclusion of graded connected commutative algebras over $\mathbb{F}$, and*

(1) *H'^* is an algebra over $\mathcal{P}^*$,*

(2) *for every $h'' \in \mathrm{H}''^*$ we have that $h''^p \in C(\mathrm{H}'^*)$,*

(3) *every $c \in C(\mathrm{H}'^*)$ has a p-th root in H''^*, and*

(4) *H''^* is reduced.*

Then there is a unique extension of the $\mathcal{P}^*$-action on H'^* to H''^*.

Proof: We define an action of the Steenrod reduced power operations in the following way[3]: For $h'' \in \mathrm{H}''^*$ and $j \in \mathbb{N}_0$ the equation

$$\left(\mathcal{P}^j(h'')\right)^p := \mathcal{P}^{jp}(h''^p) \quad \forall\ j$$

uniquely defines $\mathcal{P}^j(h'') \in \mathrm{H}''^*$, by (3), since, $h''^p \in C(\mathrm{H}'^*)$ by (2), $\mathcal{P}^{jp}(h''^p) \in C(\mathrm{H}'^*)$ by Lemma 4.1.2, and raising to the p-th power is monic in H''^* by (4). Next observe that this action is additive and satisfies the Cartan formulae: the proof on page 37 for the case of fields can be copied word-for-word. Finally we need to verify that the Adem-Wu relations evaluate to zero on H''^*. We do this by employing the Bullett-Macdonald identity instead. So let $h'' \in \mathrm{H}''^*$. Then

$$\begin{aligned}
&\left(\left(\boldsymbol{P}(s)\boldsymbol{P}(1) - \boldsymbol{P}(u)\boldsymbol{P}(t^q)\right)(h'')\right)^p \\
&= \left(\boldsymbol{P}(s)\boldsymbol{P}(1)(h'')\right)^p - \left(\boldsymbol{P}(u)\boldsymbol{P}(t^q)(h'')\right)^p \\
&= \boldsymbol{P}(s)\left(\boldsymbol{P}(1)(h'')\right)^p - \boldsymbol{P}(u)\left(\boldsymbol{P}(t^q)(h'')\right)^p \\
&= \boldsymbol{P}(s)\boldsymbol{P}(1)(h''^p) - \boldsymbol{P}(u)\boldsymbol{P}(t^q)(h''^p) \\
&= \left(\boldsymbol{P}(s)\boldsymbol{P}(1) - \boldsymbol{P}(u)\boldsymbol{P}(t^q)\right)(h''^p) \\
&= 0,
\end{aligned}$$

since $h''^p \in \mathrm{H}'^*$ and H'^* is an algebra over $\mathcal{P}^*$ by assumption (1). Therefore, since raising te the p-th power is monic in H''^* by (4),

$$\left(\boldsymbol{P}(s)\boldsymbol{P}(1) - \boldsymbol{P}(u)\boldsymbol{P}(t^q)\right)(h'') = 0$$

for every $h'' \in \mathrm{H}''^*$. •

If $\mathrm{H}_1^* = \mathrm{H}^*$ then H^* was already $\mathcal{P}^*$-inseparably closed. If $C(\mathrm{H}_1^*) = \mathrm{H}_1^{*p}$ then H_1^* is $\mathcal{P}^*$-inseparably closed. Otherwise we repeat this construction: take the subalgebra $C_1^* := C(\mathrm{H}_1^*)$ of $\mathcal{P}^{\Delta_i}$-constants in H_1^* and find a set $\mathcal{S}_1$ of generators as a module over $(\mathrm{H}_1^*)^p$. Then form the algebra

$$\mathrm{H}_2^* := \left(\mathrm{H}_1^* \otimes_{\mathbb{F}} \mathbb{F}[\gamma_s \mid s \in \mathcal{S}_1]\right) / \mathcal{R}ad(\gamma_s^p - s \mid s \in \mathcal{S}_1).$$

Again, if $\mathrm{H}_2^{*p} = C_2^*$ stop, otherwise repeat. This way we get an ascending chain of algebras over the Steenrod algebra

$$\mathrm{H}^* / \mathcal{N}il(\mathrm{H}^*) =: \mathrm{H}_0{}^* \subseteq \mathrm{H}_1^* \subseteq \cdots \subseteq \mathrm{H}_k^* \subseteq \cdots,$$

each extension being integral. If H_0^* is Noetherian then H_i^* is also Noetherian for all $i \geq 0$ and the extensions are finite, by an iterated use of Lemmata 4.1.1 and 4.1.3 (2).

---

[3] Recall from Section 2.3 the analogous construction on the field level.

PROPOSITION 4.1.5: *Let* H^* *be reduced. The colimit of this chain of algebras is the* $\mathcal{P}^*$*-inseparable closure of* H^*.

PROOF: First we need to show that $\mathrm{colim}(\mathrm{H}_i^*)$ is $\mathcal{P}^*$-inseparably closed. To see this suppose $c \in C(\mathrm{colim}(\mathrm{H}_i^*))$ is a $\mathcal{P}^{\Delta_i}$-constant. Then there exists an index i_0 such that $c \in \mathrm{H}_{i_0}^*$, because our colimit is filtered. So c has a p-th root in $\mathrm{H}_{i_0+1}^* \subseteq \mathrm{colim}(\mathrm{H}_i^*)$ as claimed.

Next we show that $\mathrm{colim}(\mathrm{H}_i^*)$ is indeed the $\mathcal{P}^*$-inseparable closure $\sqrt[\mathcal{P}^*]{\mathrm{H}^*}$. For that let $f : \mathrm{H}^* \hookrightarrow \mathrm{H}'^*$ be an inclusion of algebras over $\mathcal{P}^*$, and let H'^* be $\mathcal{P}^*$-inseparably closed. We show by induction[4] on i that f factorizes through the canonical map $\psi_i : \mathrm{H}^* \hookrightarrow \mathrm{H}_i^*$.

If $i = 1$ then for every element $h \in \mathrm{H}_1^*$ we have that $h^p \in \mathrm{H}^*$. Since

$$\mathcal{P}^{\Delta_i}\left(f(h^p)\right) = f\left(\mathcal{P}^{\Delta_i}(h^p)\right) = 0 \in \mathrm{H}'^*$$

the image $f(h^p)$ has a p-th root, say $\overline{h}$, in H'^*. Then $f_1(h) := \overline{h}$ defines the required inclusion

$$f : \mathrm{H}^* \underset{\psi_1}{\hookrightarrow} \mathrm{H}_1^* \underset{f_1}{\hookrightarrow} \mathrm{H}'^*.$$

If $i > 1$ then by induction we have

$$f : \mathrm{H}^* \underset{\psi_{i-1}}{\hookrightarrow} \mathrm{H}_{i-1}^* \underset{f_{i-1}}{\hookrightarrow} \mathrm{H}'^*.$$

Since $\mathrm{H}_i^* = (\mathrm{H}_{i-1}^*)_1$ we are done.

Finally, we define

$$\overline{f} : \mathrm{colim}(\mathrm{H}_i{}^*) \hookrightarrow \mathrm{H}'^*,\ \overline{f}(h) = f_{i_0}(h),$$

where i_0 is such that $h \in \mathrm{H}_{i_0}^*$. By construction this is a well defined map of algebras over $\mathcal{P}^*$, such that

$$f : \mathrm{H}^* \underset{\psi}{\hookrightarrow} \mathrm{colim}(\mathrm{H}_i^*) \underset{\overline{f}}{\hookrightarrow} \mathrm{H}'^*.$$

•

---

[4] We could equally proceed in the following way: if $h \in \mathrm{colim}(\mathrm{H}_i^*)$ then $h^{p^t} \in \mathrm{H}^*$ for some large enough t. From the viewpoint of the Steenrod algebra the element h^{p^t} is a p^t-th power, because the operations $\mathcal{P}^{p^r\Delta_i}$, for $r \leq t$, detect p^t-th powers, see Lemma 1.2 in [34], and f commutes with the action of the Steenrod algebra. Since H'^* is assumed to be $\mathcal{P}^*$-inseparably closed it has a unique p^t-th root $\overline{h} \in \mathrm{H}'^*$. So, defining $\overline{f}(h) := \overline{h}$ leads to the required factorization.

REMARK: A posteriori we see that the construction of H_1^* on page 57 does not depend on the choice of the set $\mathcal{S}$ of module generators of C^* over $(\mathrm{H}^*)^p$: Parts (3) and (4) of Lemma 4.1.3 tell us that

$$\mathrm{H}_1^* = \left\{h \in \sqrt[\mathcal{P}^*]{\mathrm{H}^*} \mid h^p \in C^*\right\}.$$

This makes $(-)_1$ into functor from the category of reduced graded connected commutative $\mathbb{F}$-algebras over the Steenrod algebra to itself by Lemmata 4.1.3 and 4.1.4. Finally note that if

$$\varphi : \mathrm{H'}^* \longrightarrow \mathrm{H''}^*$$

is a map in our category then the proof of Proposition 4.1.5 tells us how to extend the map φ to $\varphi_1 : \mathrm{H'}_1^* \longrightarrow \mathrm{H''}_1^*$, to wit:

$$\varphi_1(h') = h'' \quad \forall\, h' \in \mathrm{H'}_1^*,$$

where $h'' \in \mathrm{H''}_1^*$ is uniquely given by

$$(h'')^p = \varphi\left((h')^p\right).$$

We close this section with an example of a $\mathcal{P}^*$-inseparably closed algebra which has nilpotent elements, just to make clear that, just because we couldn't construct a $\mathcal{P}^*$-inseparable closure in general an algebra can well be $\mathcal{P}^*$-inseparably closed, or have a $\mathcal{P}^*$-inseparable closure.

EXAMPLE 1: Take a polynomial ring in one indeterminat of degree 1 and mod out the ideal which is generated by the p^2-th power of the generator

$$\sqrt[\mathcal{P}^*]{\mathrm{H}^*} = \mathbb{F}[x]/(x^{p^2}).$$

This is certainly $\mathcal{P}^*$-inseparably closed, in particular its the $\mathcal{P}^*$-inseparable closure of

$$\mathrm{H}^* = \mathbb{F}[x^p]/(x^{p^2}).$$

4.2 The Animal and its Properties

In this section we are going to establish the main properties of $\sqrt[\mathcal{P}^*]{\mathrm{H}^*}$, and add some examples. We do not assume that H^* is Noetherian. However, we need to assume that H^* is reduced, because otherwise our construction method would fail, recall Lemma 4.1.4. Let's start with the basics:

PROPOSITION 4.2.1: *Let H^* be a reduced algebra over the Steenrod algebra. Its $\mathcal{P}^*$-inseparable closure $\sqrt[\mathcal{P}^*]{\mathrm{H}^*}$ has the following properties:*

(1) *$\sqrt[\mathcal{P}^*]{\mathrm{H}^*}$ is a graded, connected, commutative algebra over $\mathbb{F}$.*

(2) *$\mathcal{N}il(\mathrm{H}_i^*) = (0)$, $\forall\, i$, and $\mathcal{N}il(\sqrt[\mathcal{P}^*]{\mathrm{H}^*}) = (0)$.*

(3) *Consider* $\mathrm{H}^* = \mathrm{H}_0^* \subseteq \ldots \subseteq \mathrm{H}_i^* \subseteq \ldots \subseteq \sqrt[\mathcal{P}^*]{\mathrm{H}^*}$. *If one of the algebras in this chain is an integral domain, then all the others are integral domains too.*

(4) $\mathrm{H}^* \hookrightarrow \sqrt[\mathcal{P}^*]{\mathrm{H}^*}$ *is an integral extension and* $\dim(\sqrt[\mathcal{P}^*]{\mathrm{H}^*}) = \dim(\mathrm{H}^*)$.

(5) *If* H^* *is an integrally closed domain then so is* $\sqrt[\mathcal{P}^*]{\mathrm{H}^*}$.

(6) *If* H^* *has an unstable action of the Steenrod algebra, then so does* $\sqrt[\mathcal{P}^*]{\mathrm{H}^*}$.

PROOF:

AD (1) : The first statement is true by construction, because filtered colimits exist in the category of graded connected commutative algebras over a field $\mathbb{F}$, see e.g. [16] Appendix A 6.3.

AD (2) : Since H^* is reduced we use Lemma 4.1.3 to conclude inductively that H_i^* is reduced for all $i \geq 0$. If $h \in \mathcal{N}il(\sqrt[\mathcal{P}^*]{\mathrm{H}^*}) \subset \sqrt[\mathcal{P}^*]{\mathrm{H}^*}$ then by construction there exists an $i_0 \in \mathbb{N}_0$ such that $h \in \mathrm{H}_{i_0}^*$, i.e., $h \in \mathcal{N}il(\mathrm{H}_{i_0}^*) = (0)$.

AD (3) : If $\sqrt[\mathcal{P}^*]{\mathrm{H}^*}$ is an integral domain, then so are its subalgebras H_i^*, for all $i \geq 0$ and hence so is H^*.

Conversely, if H^* is an integral domain, let $h,\ h' \in \sqrt[\mathcal{P}^*]{\mathrm{H}^*}$ with $hh' = 0$. Then for a suitably large $l \in \mathbb{N}_0$ we have

$$h^{p^l},\ h'^{p^l} \in \mathrm{H}^* \text{ and } h^{p^l} h'^{p^l} = (hh')^{p^l} = 0.$$

Since H^* has no zero divisors, h or h' is zero.

AD (4) : In the preceding section we have found that the $\mathcal{P}^*$-inseparable closure $\sqrt[\mathcal{P}^*]{\mathrm{H}^*}$ of H^* is just a filtered colimit of a certain chain of algebras

$$\mathrm{H}^* = \mathrm{H}_0^* \subseteq \mathrm{H}_1^* \subseteq \cdots \subseteq \mathrm{H}_i^* \subseteq \cdots,$$

where each extension is integral. Hence the extension

$$\mathrm{H}^* \hookrightarrow \sqrt[\mathcal{P}^*]{\mathrm{H}^*}$$

is integral, and therefore

$$\dim(\sqrt[\mathcal{P}^*]{\mathrm{H}^*}) = \dim(\mathrm{H}^*).$$

AD (5) : By (3) $\sqrt[\mathcal{P}^*]{\mathrm{H}^*}$ is an integral domain because by assumption H^* is. Consider the diagram

$$\begin{array}{ccccc}
\mathrm{H}^* = \overline{\mathrm{H}^*} & \underset{\text{integral}}{\hookrightarrow} & \sqrt[\mathcal{P}^*]{\mathrm{H}^*} & \underset{\text{integral}}{\hookrightarrow} & \overline{\sqrt[\mathcal{P}^*]{\mathrm{H}^*}} \\
\downarrow & & \downarrow & & \downarrow \\
FF(\mathrm{H}^*) & \hookrightarrow & FF(\sqrt[\mathcal{P}^*]{\mathrm{H}^*}) & = & FF\left(\overline{\sqrt[\mathcal{P}^*]{\mathrm{H}^*}}\right).
\end{array}$$

Let $\frac{h_1}{h_2} \in FF(\sqrt[\mathcal{P}^*]{\mathrm{H}^*})$ be integral over $\sqrt[\mathcal{P}^*]{\mathrm{H}^*}$. Since $h_1, h_2 \in \sqrt[\mathcal{P}^*]{\mathrm{H}^*}$ there is an $l \in \mathbb{N}_0$ such that

$$h_1^{p^l},\ h_2^{p^l} \in \mathrm{H}^*$$

and hence

$$\frac{h_1^{p^l}}{h_2^{p^l}} \in FF(\mathrm{H}^*).$$

Moreover, by transitivity, $\frac{h_1}{h_2}$ is integral over H^*, hence $\left(\frac{h_1}{h_2}\right)^{p^l}$ is integral over H^*, and so

$$\left(\frac{h_1}{h_2}\right)^{p^l} \in \mathrm{H}^*.$$

Therefore

$$\frac{h_1}{h_2} \in \sqrt[\mathcal{P}^*]{\mathrm{H}^*}$$

as we claimed.

AD (6) : If H^* is an unstable algebra, so is its subalgebra $C^* = C(\mathrm{H}^*)$. Then the definition of the $\mathcal{P}^*$-action on H_1^* given in the proof of Lemma 4.1.4 is unstable. This may be seen as follows: with the notation of Lemma 4.1.3

$$\gamma_s^p = s \quad \text{and} \quad \left(\mathcal{P}^j(\gamma_s)\right)^p := \mathcal{P}^{jp}(s)$$

for $s \in \mathcal{S}$. Hence

$$\begin{aligned}
\left(\mathcal{P}^{\deg(\gamma_s)}(\gamma_s) - \gamma_s^p\right)^p &= \left(\mathcal{P}^{\deg(\gamma_s)}(\gamma_s) - s\right)^p \\
&= \left(\mathcal{P}^{\deg(\gamma_s)}(\gamma_s)\right)^p - s^p \\
&= \mathcal{P}^{\deg(\gamma_s)p}(\gamma_s^p) - \mathcal{P}^{\deg(\gamma_s)p}(s) \\
&= \mathcal{P}^{\deg(\gamma_s)p}(\gamma_s^p - s) \\
&= \mathcal{P}^{\deg(s)}(s - s) \\
&= 0,
\end{aligned}$$

and therefore, because H_1^* is reduced,

$$\mathcal{P}^{\deg(\gamma_s)}(\gamma_s) - \gamma_s^p = 0,$$

for every $s \in \mathcal{S}$. Equally, we see that

$$\mathcal{P}^j(\gamma_s) = 0 \quad \forall\, j > \deg(\gamma_s),$$

for every $s \in \mathcal{S}$. So the action of the Steenrod algebra on H_1^* is unstable. Since by part (2) the algebra H_1^* is again reduced we can proceed to get inductively that all involved algebras H_i^*, $i = 0, 1, \ldots$, are reduced and carry an unstable Steenrod algebra action. Finally note that an element h in the colimit $\sqrt[\mathcal{P}^*]{\mathrm{H}^*}$ is contained in some H_i^*, hence is unstable and we are done. •

Remark: Statement (5) in the above Proposition 4.2.1 can be generalized to algebras with zero divisors. One would have to consider the total *ring* of fractions[5] instead of the *field* of fractions and define integrally closed as being integrally closed in its total ring of fractions.

Statement (4) of the preceding proposition does *not* imply that $\sqrt[\mathcal{P}^*]{\mathrm{H}^*}$ is Noetherian whenever H^* is. A ring can have finite Krull dimension without being Noetherian: Let $x_1, x_2, \ldots$ be linear forms, and consider the algebra

$$\mathbb{F}[x_1, x_2, \ldots]/\left(x_2^q, x_3^q, \ldots\right)$$

It is not Noetherian, because the maximal ideal is not finitely generated. However, its dimension is finite, indeed one, because

$$\left(x_2^q, x_3^q, \ldots\right) \subsetneqq \left(x_1, x_2^q, x_3^q, \ldots\right)$$

is a maximal saturated chain of prime ideals. Also, that a ring extension, where the smaller ring is Noetherian, is integral, doesn't guarantees Noetherianess for the bigger ring, see Example 1 in Section 6.1. Indeed, the question, whether $\sqrt[\mathcal{P}^*]{\mathrm{H}^*}$ is Noetherian, is difficult. We will answer it in Sections 6.1 and 6.3. The reverse problem, namely assuming that $\sqrt[\mathcal{P}^*]{\mathrm{H}^*}$ is Noetherian and asking what can we say about H^* is a lot easier, as the following discussion shows.

The following lemma was suggested by Larry Smith, [38].

Lemma 4.2.2: *Consider an increasing chain of graded connected commutative* $\mathbb{F}$*-algebras*

$$\mathrm{H}_0^* \subseteq \mathrm{H}_1^* \subseteq \mathrm{H}_2^* \subseteq \cdots.$$

If their colimit $\mathrm{colim}(\mathrm{H}_i^*)$ *is Noetherian, then there exists a* $r \in \mathbb{N}_0$ *such that*

$$\mathrm{H}_r^* = \mathrm{H}_{r+1}^* = \cdots = \mathrm{colim}(\mathrm{H}_i^*).$$

Proof: Choose generators $h_1, \ldots, h_m$ for $\mathrm{colim}(\mathrm{H}_i^*)$ as an $\mathbb{F}$-algebra. For each $i = 1, \ldots, m$ we can find a $r_i \in \mathbb{N}_0$ such that $h_i \in \mathrm{H}_{r_i}^*$. Set

$$r := \max\{r_1, \ldots, r_m\}.$$

Then

$$\mathrm{colim}(\mathrm{H}_i^*) \subseteq \mathrm{H}_r^* \subseteq \mathrm{H}_{r+1}^* \subseteq \cdots \subseteq \mathrm{colim}(\mathrm{H}_i^*)$$

and we are done. •

---

[5] For a ring H^* the **total ring of fractions** $RF(\mathrm{H}^*)$ is defined to be an over ring of H^* such that

(1) $RF(\mathrm{H}^*)$ has an identity,

(2) every element of $RF(\mathrm{H}^*)$ has the form $\frac{h_1}{h_2}$, where $h_1, h_2 \in \mathrm{H}^*$ and in addition h_2 is not a zero divisor, and

(3) every non-zero divisor of H^* has an inverse in $RF(\mathrm{H}^*)$,

see e.g. §19 of [48]. Again, in our work we, of course, consider only homogeneous elements $h_1, h_2, \ldots$, compare the remarks at the beginning of Section 2.1.

PROPOSITION 4.2.3: *Let* H^* *be a reduced algebra over the Steenrod algebra. If* $\sqrt[\mathcal{P}^*]{\mathrm{H}^*}$ *is Noetherian then so is* H^*.

PROOF: Choose a homogeneous system of parameters $h_1, \ldots, h_n$ for $\sqrt[\mathcal{P}^*]{\mathrm{H}^*}$. Then there exist indices r_j such that

$$h_j \in \mathrm{H}^*_{r_j} \quad \text{for } j = 1, \ldots, n.$$

Set

$$r := \max\{r_1, \ldots, r_n\}$$

then we have

$$h_1, \ldots, h_n \in \mathrm{H}^*_r$$

and by construction

$$h_1^{p^r}, \ldots, h_n^{p^r} \in \mathrm{H}^*$$

which is still a system of parameters for $\sqrt[\mathcal{P}^*]{\mathrm{H}^*}$. Hence the $\mathcal{P}^*$-inseparable closure, $\sqrt[\mathcal{P}^*]{\mathrm{H}^*}$, is a finite module over $\mathbb{F}[h_1^{p^r}, \ldots, h_n^{p^r}]$, and therefore so is H^*. Therefore H^* is Noetherian. •

ALTERNATIVE PROOF: If $\sqrt[\mathcal{P}^*]{\mathrm{H}^*}$ is Noetherian, then the chain

$$\mathrm{H}^* = \mathrm{H}^*_0 \subseteq \mathrm{H}^*_1 \subseteq \cdots \subseteq \mathrm{H}^*_r = \sqrt[\mathcal{P}^*]{\mathrm{H}^*}$$

becomes stationary after, say r steps, by Lemma 4.2.2. Choose a set $h_1, \ldots, h_m$ of algebra generators for $\sqrt[\mathcal{P}^*]{\mathrm{H}^*}$. Then, by definition of the inseparable closure we have [6]

$$\mathbb{F} < h_1^{p^r}, \ldots, h_m^{p^r} > \subseteq \mathrm{H}^* \subseteq \sqrt[\mathcal{P}^*]{\mathrm{H}^*} = \mathbb{F} < h_1, \ldots, h_m > .$$

So, $\sqrt[\mathcal{P}^*]{\mathrm{H}^*}$ is a finitely generated $\mathbb{F} < h_1^{p^r}, \ldots, h_m^{p^r} >$-module, hence so is H^*. Therefore H^* is Noetherian. •

The following characterization of the case when $\sqrt[\mathcal{P}^*]{\mathrm{H}^*}$ is Noetherian will be useful in later chapters.

PROPOSITION 4.2.4 : *Let* H^* *be a reduced Noetherian algebra over the Steenrod algebra. Then the following two statements are equivalent:*

(1) $\sqrt[\mathcal{P}^*]{\mathrm{H}^*}$ *is Noetherian.*

(2) *The chain of algebras*

$$\mathrm{H}^* \overset{\varphi_0}{\hookrightarrow} \mathrm{H}^*_1 \overset{\varphi_1}{\hookrightarrow} \mathrm{H}^*_2 \overset{\varphi_2}{\hookrightarrow} \cdots$$

becomes stationary after a finite number, say r *steps, so that their colimit is a finite union*

$$\operatorname{colim}\,(\mathrm{H}^*_i) = \bigcup_{i=0}^{r} \mathrm{H}^*_i.$$

---

[6] The notation $\sqrt[\mathcal{P}^*]{\mathrm{H}^*} = \mathbb{F}\langle h_1, \ldots, h_m\rangle$ means that $\sqrt[\mathcal{P}^*]{\mathrm{H}^*}$ is an $\mathbb{F}$-algebra generated by the elements $h_1, \ldots, h_m$; while $\mathbb{F}[h_1, \ldots, h_m]$ denotes a *polynomial* algebra over $\mathbb{F}$ generated by $h_1, \ldots, h_m$.

If in addition H^* *is an integral domain then the preceding statements are equivalent to*

(3) *The chain of fields of fractions*

$$FF(\mathrm{H}^*) \hookrightarrow FF(\mathrm{H}_1^*) \hookrightarrow FF(\mathrm{H}_2^*) \hookrightarrow \cdots$$

becomes stationary after a finite number, say s *steps, so that their colimit is a finite union*

$$\operatorname{colim}\left(FF(\mathrm{H}_i^*)\right) = \bigcup_{i=0}^{s} FF(\mathrm{H}_i^*).$$

Moreover $r \geq s$.

Proof: We start by proving the equivalence of (1) and (2).

AD (1) $\Rightarrow$ (2) : This is easy. Apply Lemma 4.2.2.

AD (2) $\Rightarrow$ (1) : This is equally easy. Just observe that

$$\sqrt[\mathcal{P}^*]{\mathrm{H}^*} := \operatorname{colim}(\mathrm{H}_i^*) = \bigcup_{i=0}^{r} \mathrm{H}_i^* = \mathrm{H}_r^*.$$

Since H_r^* is finite as an H^*-module by construction, and H^* is Noetherian by assumption, we conclude that $\sqrt[\mathcal{P}^*]{\mathrm{H}^*} = \mathrm{H}_r^*$ is Noetherian.

Now assume that H^* is an integral domain. We prove the implications (2) $\Rightarrow$ (3) $\Rightarrow$ (1).

AD (2) $\Rightarrow$ (3) : This is trivial.

AD (3) $\Rightarrow$ (1) : We have the following diagram

$$\begin{array}{ccccc} \mathrm{H}^* & \underset{\text{integral}}{\hookrightarrow} & \sqrt[\mathcal{P}^*]{\mathrm{H}^*} & \hookrightarrow & \overline{\mathrm{H}^*}_{FF(\sqrt[\mathcal{P}^*]{\mathrm{H}})} \\ \downarrow & & \downarrow & & \\ FF(\mathrm{H}^*) & \hookrightarrow & FF(\sqrt[\mathcal{P}^*]{\mathrm{H}^*}) & & \end{array}$$

where by assumption the field extension $FF(\mathrm{H}^*) \hookrightarrow FF(\sqrt[\mathcal{P}^*]{\mathrm{H}^*})$ is finite. Hence by an oldie from Emmy, see e.g. [16] Corollary 13.13, the integral closure of H^* in the bigger field, denoted by $\overline{\mathrm{H}^*}_{FF(\sqrt[\mathcal{P}^*]{\mathrm{H}})}$, is again Noetherian, and a finitely generated H^*-module. So $\sqrt[\mathcal{P}^*]{\mathrm{H}^*}$ is Noetherian.

Finally, note that if

$$\mathrm{H}_r^* = \mathrm{H}_{r+1}^* = \cdots$$

then,

$$FF(\mathrm{H}_r^*) = FF(\mathrm{H}_{r+1}^*) = \cdots$$

hence $r \geq s$. •

See Example 1 of Section 7.4 for an example where $r > s$ actually occurs.

We proceed to investigate the relation between the $\mathcal{P}^*$-inseparable closure on the algebra and field levels. This will be of use in later chapters.

LEMMA 4.2.5: *Let* H^* *be an integral domain. Then the field extension*

$$FF(\mathrm{H}^*) \hookrightarrow FF(\sqrt[\mathcal{P}^*]{\mathrm{H}^*})$$

is $\mathcal{P}^*$*-purely inseparable.*

PROOF: By Proposition 4.2.1 (3) the $\mathcal{P}^*$-inseparable closure $\sqrt[\mathcal{P}^*]{\mathrm{H}^*}$ is also an integral domain, so we can apply the field of fraction functor. The field extension is algebraic, since the ring extension is integral. However, for every

$$\frac{h_1}{h_2} \in FF(\sqrt[\mathcal{P}^*]{\mathrm{H}^*}) \text{ with } h_1,\ h_2 \in \sqrt[\mathcal{P}^*]{\mathrm{H}^*}$$

there is an l such that

$$h_1^{p^l},\ h_2^{p^l} \in \mathrm{H}^*$$

hence

$$\frac{h_1^{p^l}}{h_2^{p^l}} \in FF(\mathrm{H}^*),$$

i.e., the field extension is $*$-purely inseparable. Since both fields inherit an action of the Steenrod algebra from H^*, resp. $\sqrt[\mathcal{P}^*]{\mathrm{H}^*}$, and the various actions are compatible, we remain in our category and the extension is $\mathcal{P}^*$-purely inseparable. •

The following proposition shows that in the case of integral domains our notion of the $\mathcal{P}^*$-inseparable closure of an algebra fits naturally with the notion of the $\mathcal{P}^*$-inseparable closure of its field of fractions.

PROPOSITION 4.2.6: *Let* H^* *be an integral domain. The field of fractions of its* $\mathcal{P}^*$*-inseparable closure is the* $\mathcal{P}^*$*-inseparable closure of its field of fractions, i.e.,*

$$FF(\sqrt[\mathcal{P}^*]{\mathrm{H}^*}) = \big(FF(\mathrm{H}^*)\big)_{\mathcal{P}^*\text{-insep}}.$$

PROOF: Recall our construction method for $\sqrt[\mathcal{P}^*]{\mathrm{H}^*}$. It leads to the following diagram

$$\begin{array}{ccccc}
\mathrm{H}^* = \mathrm{H}_0^* & \hookrightarrow & \mathrm{H}_1^* & \hookrightarrow \cdots \hookrightarrow & \operatorname{colim}(\mathrm{H}_i^*) \\
\downarrow & & \downarrow & & \downarrow \\
FF(\mathrm{H}^*) & \hookrightarrow & FF(\mathrm{H}_1^*) & \hookrightarrow \cdots \hookrightarrow & FF\big(\operatorname{colim}(\mathrm{H}_i^*)\big)
\end{array}$$

where we have $\mathcal{P}^*$-purely inseparable field extensions by construction. So certainly

$$FF\big(\operatorname{colim}(\mathrm{H}_i^*)\big) \subseteq \big(FF(\mathrm{H}^*)\big)_{\mathcal{P}^*\text{-insep}}.$$

Moreover, we could consider the colimit, $\operatorname{colim}(FF(\mathrm{H}_i^*))$, of the sequence[7]

$$FF(\mathrm{H}^*) \hookrightarrow FF(\mathrm{H}_1^*) \hookrightarrow FF(\mathrm{H}_2^*) \hookrightarrow \cdots.$$

[7] The filtered colimit of graded fields is a graded field, compare [16] appendix A 6.3.

We have

$$\operatorname{colim}\left(FF(\mathrm{H}_i^*)\right) \subseteq FF\left(\operatorname{colim}(\mathrm{H}_i^*)\right),$$

because every $h \in \operatorname{colim}\left(FF(\mathrm{H}_i^*)\right)$ is by definition contained in some $FF(\mathrm{H}_{i_0}^*)$ which in turn is contained in $FF\left(\operatorname{colim}(\mathrm{H}_i^*)\right)$, i.e., we have field extensions

$$FF(\mathrm{H}^*) \subseteq \operatorname{colim}\left(FF(\mathrm{H}_i^*)\right) \subseteq FF\left(\operatorname{colim}(\mathrm{H}_i^*)\right) \subseteq FF\left(\mathrm{H}^*\right)_{\mathcal{P}^*\text{-insep}}.$$

Let $h \in FF\left(\mathrm{H}^*\right)_{\mathcal{P}^*\text{-insep}}$, then there is an l such that

$$h^{p^l} \in FF(\mathrm{H}^*)$$

and if we choose l to be minimal with this property we get

$$h \in FF(\mathrm{H}^*)_l = FF(\mathrm{H}_l^*) \subseteq \operatorname{colim}\left(FF(\mathrm{H}_i^*)\right).$$

Therefore the three fields are the same

$$\operatorname{colim}\left(FF(\mathrm{H}_i^*)\right) = FF\left(\operatorname{colim}(\mathrm{H}_i^*)\right) = FF\left(\mathrm{H}^*\right)_{\mathcal{P}^*\text{insep}}$$

and we are done. •

There is an obvious corollary to this.

COROLLARY 4.2.7: *If* H^* *is a* $\mathcal{P}^*$*-inseparably closed integral domain then its field of fractions is also* $\mathcal{P}^*$*-inseparably closed. If in addition* H^* *is integrally closed then these two statements are equivalent.*

PROOF: If H^* is $\mathcal{P}^*$-inseparably closed then by Proposition 4.2.6

$$FF(\mathrm{H}^*) = FF\left(\operatorname{colim}(\mathrm{H}_i^*)\right) = \operatorname{colim}\left(FF(\mathrm{H}_i^*)\right) = FF\left(\mathrm{H}^*\right)_{\mathcal{P}^*\text{-insep}}.$$

In order to prove the second statement assume that H^* is integrally closed and that the field of fractions of H^* is $\mathcal{P}^*$-inseparably closed. Then the above proposition again gives that

$$FF\left(\operatorname{colim}(\mathrm{H}_i^*)\right) = \operatorname{colim}\left(FF(\mathrm{H}_i^*)\right) = FF\left(\mathrm{H}^*\right)_{\mathcal{P}^*\text{-insep}} = FF(\mathrm{H}^*).$$

So, by Proposition 4.2.1 (3), (4) and (5)

$$\mathrm{H}^* \hookrightarrow \operatorname{colim}(\mathrm{H}_i^*)$$

is an integral extension of integrally closed integral domains with the same field of fractions. Therefore they are equal. •

REMARK: If, on the other hand, we start with a $\mathcal{P}^*$-inseparably closed field $\mathbb{K}^*$ then its unstable part, $\mathcal{U}n(\mathbb{K}^*)$, is also $\mathcal{P}^*$-inseparably closed: If $k \in \sqrt[\mathcal{P}^*]{\mathcal{U}n(\mathbb{K}^*)}$ then $k \in \mathbb{K}^*$, because $\mathbb{K}^*$ is $\mathcal{P}^*$-inseparably closed, and moreover $k^{p^e} \in \mathcal{U}n(\mathbb{K}^*)$ for some suitably large $e \in \mathbb{N}_0$. However, the Cartan formulae give us

$$\mathcal{P}^r(k^{p^e}) = \left(\mathcal{P}^{\frac{r}{p^e}}(k)\right)^{p^e},$$

and so for any $r > \deg(k)p^e$ we have

$$0 = \mathcal{P}^{\frac{r}{p^e}}(k).$$

Moreover,

$$\left(\mathcal{P}^{\deg(k)}(k)\right)^{p^e} = \mathcal{P}^{\deg(k)p^e}(k^{p^e}) = k^{qp^e}.$$

This implies

$$\mathcal{P}^{\deg(k)}(k) = k^q,$$

because $\mathbb{K}^*$ is a field, it has in particular no nilpotent elements. Therefore k is also an unstable element.

COROLLARY 4.2.8 : *The algebra of polynomial functions in n variables $\mathbb{F}[x_1, \dots, x_n]$ is $\mathcal{P}^*$-inseparably closed.*

PROOF: This follows from the above Corollary 4.2.7, because the field of fractions $\mathbb{F}(x_1, \dots, x_n)$ is $\mathcal{P}^*$-inseparably closed by Corollary 2.3.4. •

PROPOSITION 4.2.9: *The Dickson algebra $\mathcal{D}^*(n)$ is $\mathcal{P}^*$-inseparably closed.*

PROOF: Let $\sqrt[\mathcal{P}^*]{\mathcal{D}^*(n)}$ be the $\mathcal{P}^*$-inseparable closure of the Dickson algebra $\mathcal{D}^*(n)$. Then we have

$$\mathcal{D}^*(n) \hookrightarrow \sqrt[\mathcal{P}^*]{\mathcal{D}^*(n)} \hookrightarrow \mathbb{F}[V],$$

because $\mathbb{F}[V]$ is $\mathcal{P}^*$-inseparably closed by Corollary 4.2.8. Take an element $h \in \sqrt[\mathcal{P}^*]{\mathcal{D}^*(n)}$. Then there exists an $l \in \mathbb{N}_0$ such that

$$h^{p^l} \in \mathcal{D}^*(n).$$

We take a suitably high power of h^{p^l} and conclude that h^{q^l} is invariant under every element $g \in \mathrm{GL}(n, \mathbb{F})$, i.e.,

$$h^{q^l} = g\left(h^{q^l}\right) = (g(h))^{q^l}.$$

Therefore, for some non-zero element $\lambda_g \in \mathbb{F}^\times$ in the ground field, we have

$$g(h) = \lambda_g h$$

where $\lambda_g^{q^l} = 1$. Since

$$\lambda_g = \lambda_g^{q^l}$$

we get that the general linear group acts trivially on h, i.e.,

$$h \in \mathcal{D}^*(n),$$

as claimed. •

Recall that a fractal $\mathcal{D}^*(n)^{q^s}$ of the Dickson algebra is

$$\mathcal{D}^*(n)^{q^s} = \mathbb{F}[\mathbf{d}_{n,0}^{q^s}, \ldots, \mathbf{d}_{n,n-1}^{q^s}].$$

For further reference we append the obvious corollary.

COROLLARY 4.2.10: *The $\mathcal{P}^*$-inseparable closure of a fractal $\mathcal{D}^*(n)^{q^s}$ of the Dickson algebra is the Dickson algebra $\mathcal{D}^*(n)$.*

PROOF: By definition, $h^{q^s} \in \mathcal{D}^*(n)^{q^s}$ for every element $h \in \mathcal{D}^*(n)$. Therefore the extension

$$\mathcal{D}^*(n)^{q^s} \hookrightarrow \mathcal{D}^*(n)$$

is $\mathcal{P}^*$-purely inseparable. Since $\mathcal{D}^*(n)$ is $\mathcal{P}^*$-inseparably closed, by Proposition 4.2.9, we are done. •

4.3 Further Properties

In this section we study the relation between the ideals of an algebra H^* and its $\mathcal{P}^*$-inseparable closure and establish some technical properties which will be useful later on.

We continue to assume that H^* is reduced.

Let's denote by $\mathcal{P}roj(H^*)$ the spectrum of homogeneous prime ideals of H^*. Recall from the preceding section the chain of algebras leading to the inseparable closure

$$H^* = H_0^* \overset{\varphi_0}{\hookrightarrow} H_1^* \overset{\varphi_1}{\hookrightarrow} H_2^* \overset{\varphi_2}{\hookrightarrow} \cdots \subset \sqrt[\mathcal{P}^*]{H^*} = \operatorname{colim}(H_i^*),$$

and that every ring extension φ_i, for all $i \in \mathbb{N}_0$, is integral. Therefore, by lying over, Theorem 3.1.16 in [4], the induced maps

$$\varphi_i^* : \mathcal{P}roj(H_{i+1}^*) \longrightarrow \mathcal{P}roj(H_i^*),$$

and

$$\varphi^* : \mathcal{P}roj\left(\sqrt[\mathcal{P}^*]{H^*}\right) \longrightarrow \mathcal{P}roj(H^*)$$

are surjective for any $i \in \mathbb{N}_0$. The following theorem shows that these maps are also injective.

THEOREM 4.3.1: *Let* H^* *be a reduced algebra over the Steenrod algebra. With the preceding notation we have that*

$$\varphi_i^* : \mathcal{P}roj(H_{i+1}^*) \longrightarrow \mathcal{P}roj(H_i^*)$$

is a bijection for any $i \in \mathbb{N}_0$. *Moreover, this also holds for the map*

$$\varphi^* : \mathcal{P}roj\left(\sqrt[\mathcal{P}^*]{H^*}\right) \longrightarrow \mathcal{P}roj(H^*)$$

induced by the inclusion $\varphi : H^* \hookrightarrow \sqrt[\mathcal{P}^*]{H^*}$.

PROOF: Consider our map

$$\begin{array}{rccc} \varphi_i^* : & \mathcal{P}roj(H_{i+1}^*) & \longrightarrow & \mathcal{P}roj(H_i^*) \\ & \mathfrak{q} & \longmapsto & \mathfrak{p} := \mathfrak{q} \cap H_i^*. \end{array}$$

We have already seen that φ_i^* is surjective, so what's left to show that it is also injective.

Let $\mathfrak{q}_1, \mathfrak{q}_2 \in \mathcal{P}roj(H_{i+1}^*)$ such that

$$\varphi_i^*(\mathfrak{q}_1) = \mathfrak{p} = \varphi_i^*(\mathfrak{q}_2)$$

for some prime ideal $\mathfrak{p} \in \mathcal{P}roj(H_i^*)$. Denote by $\mathfrak{p}^e$ the extended ideal $(\varphi_i(\mathfrak{p})) \subset H_{i+1}^*$. Then

$$\mathfrak{p}^e = \left(\mathfrak{q}_1 \cap \mathfrak{q}_2 \cap H_i^*\right)^e \subseteq \mathfrak{q}_1 \cap \mathfrak{q}_2 \subseteq \mathfrak{q}_j \quad j = 1,\ 2.$$

Therefore

$$\mathfrak{p}^e \subseteq \mathcal{R}ad(\mathfrak{p}^e) \subseteq \mathfrak{q}_1 \cap \mathfrak{q}_2 \subseteq \mathfrak{q}_j \quad j = 1,\ 2,$$

where $\mathcal{R}ad(-)$ denotes the radical of an ideal. Hence

$$\mathfrak{p} \subseteq \mathfrak{p}^e \cap H_i^* \subseteq \mathcal{R}ad(\mathfrak{p}^e) \cap H_i^* \subseteq \mathfrak{q}_1 \cap \mathfrak{q}_2 \cap H_i^* \subseteq \mathfrak{q}_j \cap H_i^* = \mathfrak{p}, \quad j = 1,\ 2,$$

i.e., any of the ideals of this chain contracts to $\mathfrak{p}$. If we could show that the $\mathcal{R}ad(\mathfrak{p}^e) \subset H_{i+1}^*$ is a prime ideal, then we would have that

$$\mathcal{R}ad(\mathfrak{p}^e) = \mathfrak{q}_j \quad j = 1,\ 2,$$

and in particular

$$\mathfrak{q}_1 = \mathfrak{q}_2,$$

since there are no proper inclusions between the prime ideals lying over a fixed prime ideal. So, what's left to show is that

$$\mathcal{R}ad(\mathfrak{p}^e) \subset H_{i+1}^*$$

is a prime ideal. This we do by direct calculation. Let $h,\ h' \in H_{i+1}^*$ be such that their product $hh' \in \mathcal{R}ad(\mathfrak{p}^e)$.

CASE 1 : $h,\ h' \in \mathrm{H}_i^*$.
Then

$$hh' \in \mathcal{Rad}(\mathfrak{p}^e) \cap \mathrm{H}_i^* = \mathfrak{p}$$

and hence without loss of generality $h \in \mathfrak{p} \subseteq \mathcal{Rad}(\mathfrak{p}^e)$.

CASE 2 : $h \in \mathrm{H}_i^*$ and $h' \notin \mathrm{H}_i^*$.
Then $h'^p \in \mathrm{H}_i^* \setminus \{0\}$ and therefore

$$hh'^p \in \mathcal{Rad}(\mathfrak{p}^e) \cap \mathrm{H}_i^* = \mathfrak{p},$$

so $h \in \mathfrak{p} \subseteq \mathcal{Rad}(\mathfrak{p}^e)$ or $h'^p \in \mathfrak{p} \subseteq \mathfrak{p}^e$, which implies $h' \in \mathcal{Rad}(\mathfrak{p}^e)$.

CASE 3 : Neither h nor h' is in H_i^*.
Then $h^p,\ h'^p \in \mathrm{H}_{i+1}^* \setminus \{0\}$ and, as above, we conclude from

$$h^p h'^p \in \mathcal{Rad}(\mathfrak{p}^e) \cap \mathrm{H}_i^* = \mathfrak{p}$$

that, without loss of generality,

$$h^p \in \mathfrak{p} \subseteq \mathfrak{p}^e$$

and hence

$$h \in \mathcal{Rad}(\mathfrak{p}^e).$$

This proves the first part of the theorem. For the second part note that the same series of arguments as used above apply to the integral extension

$$\varphi : \mathrm{H}^* \hookrightarrow \sqrt[\mathcal{P}^*]{\mathrm{H}^*},$$

i.e., the induced map

$$\varphi^* : \mathcal{Proj}\left(\sqrt[\mathcal{P}^*]{\mathrm{H}^*}\right) \longrightarrow \mathcal{Proj}(\mathrm{H}^*)$$

is surjective and for any $\mathfrak{q}_1,\ \mathfrak{q}_2 \in \mathcal{Proj}\left(\sqrt[\mathcal{P}^*]{\mathrm{H}^*}\right)$ such that

$$\varphi^*(\mathfrak{q}_1) = \varphi^*(\mathfrak{q}_2) = \mathfrak{p} \in \mathcal{Proj}(\mathrm{H}^*),$$

we have that

$$\mathfrak{p} \subseteq \mathfrak{p}^e \cap \mathrm{H}^* \subseteq \mathcal{Rad}(\mathfrak{p}^e) \cap \mathrm{H}^* \subseteq \mathfrak{q}_1 \cap \mathfrak{q}_2 \cap \mathrm{H}^* \subseteq \mathfrak{q}_1 \cap \mathrm{H}^* = \mathfrak{p}.$$

So what is left to show is that for any prime ideal $\mathfrak{p} \subset \mathrm{H}^*$ the radical of its extension in $\sqrt[\mathcal{P}^*]{\mathrm{H}^*}$, $\mathcal{Rad}(\mathfrak{p}^e)$, is again prime.

To this end take $h,\ h' \in \sqrt[\mathcal{P}^*]{\mathrm{H}^*}$ such that $hh' \in \mathcal{Rad}(\mathfrak{p}^e)$. Then there exists an index $i_0 \in \mathbb{N}_0$ such that $h,\ h' \in \mathrm{H}_{i_0}$ and hence

$$hh' \in \mathcal{Rad}(\mathfrak{p}^e) \cap \mathrm{H}_{i_0}.$$

Therefore $h^{p^{i_0}},\ h'^{p^{i_0}} \in \mathrm{H}^*$ and hence

$$h^{p^{i_0}} h'^{p^{i_0}} \in \mathcal{Rad}(\mathfrak{p}^e) \cap \mathrm{H}^* = \mathfrak{p}.$$

Without loss of generality assume that

$$h^{p^{i_0}} \in \mathfrak{p}.$$

Let $\psi_i : \mathrm{H}^* \hookrightarrow \mathrm{H}_i^*$, for every $i \in \mathbb{N}_0$, be the canonical inclusion. We have

$$h \in \mathcal{R}ad\,(\psi_{i_0}(\mathfrak{p})) \subset \mathrm{H}_{i_0}^*$$

and hence

$$h \in \left(\mathcal{R}ad\,(\psi_{i_0}(\mathfrak{p}))\right)^e \subseteq \mathcal{R}ad\left((\psi_{i_0}(\mathfrak{p}))^e\right) = \mathcal{R}ad(\mathfrak{p}^e) \subset \sqrt[\mathcal{P}^*]{\mathrm{H}^*},$$

which was to be shown. •

Lemma 4.3.2: *Taking the $\mathcal{P}^*$-inseparable closure commutes with intersections, i.e., for two subalgebras H'^*, H''^* of a reduced algebra H^* over the Steenrod algebra one has*

$$\sqrt[\mathcal{P}^*]{\mathrm{H}'^*} \cap \sqrt[\mathcal{P}^*]{\mathrm{H}''^*} = \sqrt[\mathcal{P}^*]{\mathrm{H}'^* \cap \mathrm{H}''^*} \subseteq \sqrt[\mathcal{P}^*]{\mathrm{H}^*}.$$

Proof: Let $h \in \sqrt[\mathcal{P}^*]{\mathrm{H}'^*} \cap \sqrt[\mathcal{P}^*]{\mathrm{H}''^*}$. Then there exist $l',\ l'' \in \mathbb{N}_0$ such that

$$h^{p^{l'}} \in \mathrm{H}'^* \quad \text{and} \quad h^{p^{l''}} \in \mathrm{H}''^*.$$

Let $l = \max\{l',\ l''\}$. Then

$$h^{p^l} \in \mathrm{H}'^* \cap \mathrm{H}''^* \subseteq \sqrt[\mathcal{P}^*]{\mathrm{H}'^* \cap \mathrm{H}''^*},$$

and hence

$$h \in \sqrt[\mathcal{P}^*]{\mathrm{H}'^* \cap \mathrm{H}''^*}.$$

On the other hand let $h \in \sqrt[\mathcal{P}^*]{\mathrm{H}'^* \cap \mathrm{H}''^*}$. Then there exists an $k \in \mathbb{N}_0$ such that

$$h^{p^k} \in \mathrm{H}'^* \cap \mathrm{H}''^*,$$

and hence

$$h \in \sqrt[\mathcal{P}^*]{\mathrm{H}'^*} \cap \sqrt[\mathcal{P}^*]{\mathrm{H}''^*},$$

as we wanted. •

There is an obvious corollary.

Corollary 4.3.3: *The intersection of two $\mathcal{P}^*$-inseparably closed algebras is $\mathcal{P}^*$-inseparably closed.* •

Recall from the introduction the following definitions.

DEFINITION: *An ideal $I \subseteq \mathrm{H}^*$ has* **height** *i, if there is an isolated prime ideal associated to I with height i and i is minimal with respect to this property:*

$$\mathcal{ht}(I) := \min\{\mathcal{ht}(\mathfrak{p}) \mid I \subseteq \mathfrak{p},\ \mathfrak{p} \text{ prime}\}.$$

An ideal $I \subseteq \mathrm{H}^*$ in an algebra over the Steenrod algebra is called **$\mathcal{P}^*$-invariant** *if it is closed under the action of the Steenrod algebra*[8]. *Finally an element* $\mathbf{t} \in \mathrm{H}^*$ *is called a* **Thom class** *if the principal ideal it generates is a $\mathcal{P}^*$-invariant ideal of height one.*

We want to know what happens to $\mathcal{P}^*$-inseparably closed algebras H^*, when we take quotients by ideals $I \subset \mathrm{H}^*$. The quotient H^*/I inherits an $\mathcal{P}^*$-action if and only if I is $\mathcal{P}^*$-invariant. So, only in this context does the question make sense. However, since I might not be a radical ideal, the quotient H^*/I might have nilpotent elements, so not be reduced.

Denote by $\mathcal{Rad}(I)$ the radical of the ideal $I \subset \mathrm{H}^*$. If I is $\mathcal{P}^*$-invariant, then so is its radical by Proposition 11.2.5 in [36].

We have:

LEMMA 4.3.4: *Let H^* be a reduced algebra over the Steenrod algebra and consider the integral extension $\mathrm{H}^* \hookrightarrow \mathrm{H}_1^*$. If $\mathfrak{p} \subset \mathrm{H}^*$ is a $\mathcal{P}^*$-invariant prime ideal and $\mathfrak{q} \subset \mathrm{H}_1$ lies*[9] *over it, then*

$$\mathrm{H}_1^*/\mathfrak{q} \subseteq (\mathrm{H}^*/\mathfrak{p})_1 \subseteq (\mathrm{H}_1^*/\mathfrak{q})_1.$$

PROOF: The proof is by straightforward calculation. Take an element

$$h \in \mathrm{H}_1^*/\mathfrak{q},$$

then $h + \mathfrak{q} \subset \mathrm{H}_1^*$. Hence by construction $h^p + h'^p \subset \mathrm{H}^*$, for every $h + h' \in h + \mathfrak{q} \subset \mathrm{H}_1^*$, and therefore

$$h^p \in \mathrm{H}^*/\mathfrak{p},$$

i.e., $h \in (\mathrm{H}^*/\mathfrak{p})_1$, what proves the first inclusion. To prove the second inclusion note that

$$\mathrm{H}^* \hookrightarrow \mathrm{H}_1^*$$

implies

$$\mathrm{H}^*/\mathfrak{p} \hookrightarrow \mathrm{H}_1^*/\mathfrak{q},$$

and this in turn implies

$$(\mathrm{H}^*/\mathfrak{p})_1 \hookrightarrow (\mathrm{H}_1^*/\mathfrak{q})_1,$$

as claimed. •

---

[8] See [36] Chapter 11 and [26], [29], [33] and [35] for further results.

[9] The prime ideal $\mathfrak{q}$ is again $\mathcal{P}^*$-invariants by Theorem 2.3 in [26].

This result easily extends to

$$\mathrm{H}_1^*/\mathcal{R}ad(I) \subseteq (\mathrm{H}^*/(\mathcal{R}ad(I)\cap \mathrm{H}^*))_1 \subseteq (\mathrm{H}_1^*/\mathcal{R}ad(I))_1$$

for all $\mathcal{P}^*$-invariant ideals $I \subseteq \mathrm{H}_1^*$. Inductively we have

PROPOSITION 4.3.5: *Let* H^* *be a reduced algebra over the Steenrod algebra and consider the integral extension* $\mathrm{H}^* \hookrightarrow \sqrt[\mathcal{P}^*]{\mathrm{H}^*}$. *If* $I \subset \sqrt[\mathcal{P}^*]{\mathrm{H}^*}$ *is a* $\mathcal{P}^*$*-invariant ideal, then*

$$\mathrm{H}^*/(\mathcal{R}ad(I)\cap \mathrm{H}^*) \subseteq \sqrt[\mathcal{P}^*]{\mathrm{H}^*}/\mathcal{R}ad(I) \subseteq \sqrt[\mathcal{P}^*]{\mathrm{H}^*/(\mathcal{R}ad(I)\cap \mathrm{H}^*)} = \sqrt[\mathcal{P}^*]{\sqrt[\mathcal{P}^*]{\mathrm{H}^*}/\mathcal{R}ad(I)}.$$

•

We have another amazing result, which illustrates again the rigidity of the structure the Steenrod algebra forces on H^*.

PROPOSITION 4.3.6: *If* H^* *is* $\mathcal{P}^*$*-inseparably closed, and for any* $h \in \mathrm{H}^*$

$$\mathcal{P}^{\Delta_i}(h) = 0 \ \forall\ i \geq 0,$$

then $\mathrm{H}^* = \mathbb{F}$.

PROOF: Let $h \in \mathrm{H}^*$ be of minimal positive degree d. By assumption

$$\mathcal{P}^{\Delta_i}(h) = 0 \ \forall\ i \geq 0.$$

Since H^* is $\mathcal{P}^*$-inseparably closed h must have a p-th root in H^*, which has degree $0 < \frac{d}{p} < d$. This is a contradiction. Hence there is no element of positive degree and $\mathrm{H}^* = \mathbb{F}$, because our algebra is connected. •

REMARK: An analog statement is true if we consider fields $\mathbb{K}^*$ over $\mathcal{P}^*$, i.e., a $\mathcal{P}^*$-inseparably closed field $\mathbb{K}^*$ for which

$$\mathcal{P}^{\Delta_i}(k) = 0 \quad \forall\ i \geq 0 \quad \forall\ k \in \mathbb{K}^*$$

is concentrated in degree zero $\mathbb{K}^* = \mathbb{K}^*_{(0)}$. Its unstable part $\mathcal{U}n(\mathbb{K}^*)$ is therefore a q-Boolean algebra, because

$$k = \mathcal{P}^0(k) = k^q$$

for every $k \in \mathcal{U}n(\mathbb{K}^*) = \mathcal{U}n(\mathbb{K}^*_{(0)})$. In other words, we have

$$\mathcal{U}n(\mathbb{K}^*)^q = \mathcal{U}n(\mathbb{K}^*),$$

as in the classical situation.

Chapter 5
The Embedding Theorem I

In this chapter we are going to prove the Embedding Theorem over arbitrary Galois fields for $\mathcal{P}^*$-inseparably closed unstable algebras H^*.

Throughout the whole chapter H^* is a Δ-finite unstable integral domain.

5.1 A Dickson Algebra in $FF(\mathrm{H}^*)$

We will start this chapter with close scrutiny of the coefficients which appear in a Δ-relation.

Recall from Section 1.2 that there are non-zero elements $h_0, \ldots, h_m \in \mathrm{H}^*$ such that

$$h_0\mathcal{P}^{\Delta_0} + \cdots + h_m\mathcal{P}^{\Delta_m} = 0$$

in the H^*-module $\Delta(\mathrm{H}^*)$. If we consider the coefficients as lying in the field of fractions $FF(\mathrm{H}^*)$ of H^*, we can normalize this relation to get

$$\mathbf{t}_0\mathcal{P}^{\Delta_0} + \mathbf{t}_1\mathcal{P}^{\Delta_1} + \cdots + \mathbf{t}_{m-1}\mathcal{P}^{\Delta_{m-1}} + \mathcal{P}^{\Delta_m} = 0$$

where $\mathbf{t}_i = \frac{h_i}{h_m}$ for $i = 0, \ldots, m-1$. For convenience we define

$$\mathbf{t}_m = 1.$$

OBSERVATION 5.1.1: *Note that by minimality of $m = m(\mathrm{H}^*)$ the coefficients $\mathbf{t}_0, \ldots, \mathbf{t}_{m-1}$, $\mathbf{t}_m = 1$ are uniquely determined in $FF(\mathrm{H}^*)$, and $\deg(\mathbf{t}_i) = q^m - q^i$ by homogeneity.*

This means, that if there is another $m+1$-tuple $(a_0, \ldots, a_m) \in FF(\mathrm{H}^*)$

$$a_0\mathcal{P}^{\Delta_0} + \cdots + a_m\mathcal{P}^{\Delta_m} = 0$$

on H^*, then

$$(a_0, \ldots, a_m) = a_m(\mathbf{t}_0, \ldots, \mathbf{t}_m).$$

We are aiming to prove that these coefficients $\mathbf{t}_0, \ldots, \mathbf{t}_{m-1}$ generate a Dickson algebra inside $FF(\mathrm{H}^*)$ of Krull dimension m. First we prove that they are algebraically independent. We introduce the following convention:

CONVENTION: For $m = 0$ call

$$\mathcal{D}^*(0) := \mathbb{F}$$

the **trivial Dickson algebra**.

LEMMA 5.1.2: *The elements* $\mathbf{t}_0, \ldots, \mathbf{t}_{m-1} \in FF(\mathrm{H}^*)$, *uniquely given by the* Δ*-Theorem* 1.2.1, *are algebraically independent.*

PROOF: It is enough to show that the determinant of the generalized Jacobian matrix is not zero, see Theorem A.4.1,

$$\det\left(\mathcal{P}^{\Delta_i}(\mathbf{t}_j)\right)_{i,\, j=0,\ldots,\, m-1} \neq 0.$$

By Proposition 1.2.4

$$\begin{aligned}
0 &= \mathcal{P}^{\Delta_i}\Delta(0, \mathbf{t}_0, \ldots, \mathbf{t}_m)\\
&= \mathcal{P}^{\Delta_i}(\mathbf{t}_0)\mathcal{P}^{\Delta_0} + \cdots + \mathcal{P}^{\Delta_i}(\mathbf{t}_m)\mathcal{P}^{\Delta_m} + \mathbf{t}_0\mathcal{P}^{\Delta_i}\mathcal{P}^{\Delta_0} + \cdots + \mathbf{t}_m\mathcal{P}^{\Delta_i}\mathcal{P}^{\Delta_m}\\
&= \mathcal{P}^{\Delta_i}(\mathbf{t}_0)\mathcal{P}^{\Delta_0} + \cdots + \mathcal{P}^{\Delta_i}(\mathbf{t}_m)\mathcal{P}^{\Delta_m} + \mathbf{t}_0\mathcal{P}^{\Delta_0}\mathcal{P}^{\Delta_i} + \cdots + \mathbf{t}_m\mathcal{P}^{\Delta_m}\mathcal{P}^{\Delta_i} + \mathbf{t}_0\mathcal{P}^{\Delta_i}\\
&= \mathcal{P}^{\Delta_i}(\mathbf{t}_0)\mathcal{P}^{\Delta_0} + \cdots + \left(\mathcal{P}^{\Delta_i}(\mathbf{t}_i) + \mathbf{t}_0\right)\mathcal{P}^{\Delta_i} + \cdots + \mathcal{P}^{\Delta_i}(\mathbf{t}_{m-1})\mathcal{P}^{\Delta_{m-1}},
\end{aligned}$$

where the last two equations arise, because $\mathbf{t}_m = 1$ implies $\mathcal{P}^{\Delta_i}(\mathbf{t}_m) = 0$, and $\left(\mathbf{t}_0\mathcal{P}^{\Delta_0} + \cdots + \mathbf{t}_m\mathcal{P}^{\Delta_m}\right)\mathcal{P}^{\Delta_i} = 0$ for all i by assumption. Since m was minimal we get

$$\mathcal{P}^{\Delta_i}(\mathbf{t}_j) = \begin{cases} 0 & \text{for } i = 1, \ldots, m-1,\ i \neq j\\ -\mathbf{t}_0 & \text{for } i = j.\end{cases}$$

Therefore our matrix looks like

$$\det\left(\mathcal{P}^{\Delta_i}(\mathbf{t}_j)\right)_{i,\, j=0}^{m-1} = \det\begin{bmatrix} \mathcal{P}^{\Delta_0}(\mathbf{t}_0) & \cdots & \mathcal{P}^{\Delta_0}(\mathbf{t}_{m-1})\\ \vdots & \ddots & \vdots\\ \mathcal{P}^{\Delta_{m-1}}(\mathbf{t}_0) & \cdots & \mathcal{P}^{\Delta_{m-1}}(\mathbf{t}_{m-1})\end{bmatrix}$$

$$\begin{aligned}
&= \det\begin{bmatrix} \deg(\mathbf{t}_0)\mathbf{t}_0 & \cdots & \cdots & \cdots & \deg(\mathbf{t}_{m-1})\mathbf{t}_{m-1}\\ 0 & -\mathbf{t}_0 & 0 & \cdots & 0\\ \vdots & \ddots & \ddots & \ddots & \vdots\\ 0 & \cdots & 0 & -\mathbf{t}_0 & 0\\ 0 & \cdots & \cdots & 0 & -\mathbf{t}_0\end{bmatrix}\\
&= (-1)^{m-1}\deg(\mathbf{t}_0)\mathbf{t}_0^m\\
&= (-1)^{m-1}(q^m - 1)\mathbf{t}_0^m.
\end{aligned}$$

This determinant is non-zero if and only if $m > 0$. However, for $m = 0$ we have $\mathbf{t}_0 = \mathbf{t}_m = 1$, and, by convention, $\mathbb{F} \subseteq FF(\mathrm{H}^*)$, so we are done. •

COROLLARY 5.1.3: *The elements* $\mathbf{t}_0, \ldots, \mathbf{t}_{m-1} \in FF(\mathrm{H}^*)$, *uniquely given by the* Δ*-Theorem 1.2.1, generate a polynomial* $\mathbb{F}$*-subalgebra* [1]

$$\mathbb{F}[\mathbf{t}_0, \ldots, \mathbf{t}_{m-1}] \hookrightarrow FF(\mathrm{H}^*)$$

of $FF(\mathrm{H}^*)$. •

Of course, the Dickson algebra $\mathcal{D}^*(m)$, being a polynomial algebra of the appropriate Krull dimension, is *as algebra over* $\mathbb{F}$ isomorphic to our $\mathbb{F}[\mathbf{t}_0, \ldots, \mathbf{t}_{m-1}]$. Next, we need to show that our algebra is closed under the action of the Steenrod algebra on $FF(\mathrm{H}^*)$. For this we need some preparations.

Consider the following polynomial in $\mathrm{H}^*[X]$

$$\delta(X) := h_0 X + h_1 X^q + \cdots + h_m X^{q^m}.$$

As a polynomial over the field of fraction of H^*, $FF(\mathrm{H}^*)$, we normalize it, and get what we will call Δ**-polynomial** (recall $\mathbf{t}_m = 1$)

$$\Delta(X) := \mathbf{t}_0 X + \mathbf{t}_1 X^q + \cdots + \mathbf{t}_m X^{q^m}.$$

In the terminology of Chapter 2 $\Delta(X)$ is 1-graded. The following proposition is the optimal generalization of Lemma 5.6 in [1].

PROPOSITION 5.1.4: *Let* H^* *be a* Δ*-finite unstable integral domain with* Δ*-polynomial* $\Delta(X)$. *Let* $\mathbb{K}^*$ *be a graded field containing* H^*. *Then the following are equivalent*

(1) $\sum_{r=0}^{k} a_r \mathcal{P}^{\Delta_r} \equiv 0$ *on* H^*, *where* $a_r \in \mathbb{K}^*$ *for all* r.

(2) $\Delta(X) \mid \sum_{r=0}^{k} a_r X^{q^r}$ *in* $\mathbb{K}^*[X]$.

PROOF: If $k < m$ then

$$\sum_{r=0}^{k} a_r \mathcal{P}^{\Delta_r} = 0 \iff a_0 = \cdots = a_k = 0$$

by minimality of m and nothing needs to be shown. If $k = m$, then by Observation 5.1.1,

$$(a_0, \ldots, a_m) = a_m(\mathbf{t}_0, \ldots, \mathbf{t}_{m-1}, 1)$$

which is equivalent to

$$\Delta(X) \mid a_m \Delta(X) = \sum_{r=0}^{m} a_r X^{q^r}.$$

So we proceed by induction on k and assume that $k > m$. Then, using Lemma 1.1.8,

$$0 = \sum_{r=0}^{k} a_r \mathcal{P}^{\Delta_r} = \sum_{r=0}^{k} a_r \mathcal{P}^{\Delta_r} - a_k \sum_{i=0}^{m} \mathbf{t}_i^{q^{k-m}} \mathcal{P}^{\Delta_{k-m+i}} = \sum_{r=0}^{k-1} \tilde{a}_r \mathcal{P}^{\Delta_r}$$

[1] The term "$\mathbb{F}$-subalgebra" emphasizes that we don't know yet that this embedding is a morphism in the category of unstable algebras over the Steenrod algebra.

for certain $\tilde{a}_r \in FF(\mathrm{H}^*)$. The last equation, is, by the induction hypothesis, equivalent to

$$\Delta(X) \mid \sum_{r=0}^{k-1} \tilde{a}_r X^{q^r} = \sum_{r=0}^{k} a_r X^{q^r} - a_k \sum_{i=0}^{m} \mathbf{t}_i^{q^{k-m}} X^{q^{k-m+i}}.$$

Since,

$$\sum_{i=0}^{m} \mathbf{t}_i^{q^{k-m}} X^{q^{k-m+i}} = \left(\sum_{i=0}^{m} \mathbf{t}_i X^{q^i} \right)^{q^{k-m}} = \left(\Delta(X)\right)^{q^{k-m}},$$

we are done. •

The following lemma corresponds to Lemma 5.7 [1].

LEMMA 5.1.5 : *If* H^* *is a* Δ*-finite unstable integral domain then for any* $\alpha \geq 0$

$$\Delta(X) \mid \mathcal{P}^{\alpha} \Delta(X)$$

as polynomials[2] *over the field of fractions of* H^*, *i.e., in* $FF(\mathrm{H}^*)[X]$.

PROOF : We want to apply the above criteria for divisibility. So, lets calculate the Steenrod powers of $\Delta(X)$:

$$\begin{aligned}
\mathcal{P}^{\alpha} \Delta(X) &= \mathcal{P}^{\alpha} \left(\sum_{i=0}^{m} \mathbf{t}_i X^{q^i} \right) = \sum_{i=0}^{m} \mathcal{P}^{\alpha} \left(\mathbf{t}_i X^{q^i} \right) \\
&= \sum_{i=0}^{m} \left(\sum_{s=0}^{\alpha} \mathcal{P}^{\alpha-s}(\mathbf{t}_i) \mathcal{P}^{s}(X^{q^i}) \right) \\
&= \sum_{i=0}^{m} \left(\sum_{s=0}^{\alpha} \mathcal{P}^{\alpha-s}(\mathbf{t}_i) \left(\mathcal{P}^{\frac{s}{q^i}}(X) \right)^{q^i} \right) \\
&= \sum_{i=0}^{m} \left(\mathcal{P}^{\alpha}(\mathbf{t}_i) X^{q^i} + \mathcal{P}^{\alpha-q^i}(\mathbf{t}_i)(\mathcal{P}^1(X))^{q^i} \right) \\
&= \sum_{i=0}^{m} \left(\mathcal{P}^{\alpha}(\mathbf{t}_i) X^{q^i} + \mathcal{P}^{\alpha-q^i}(\mathbf{t}_i) X^{q^{i+1}} \right) \\
&= \sum_{i=0}^{m} \left(\mathcal{P}^{\alpha}(\mathbf{t}_i) X^{q^i} \right) + \sum_{i=0}^{m} \left(\mathcal{P}^{\alpha-q^i}(\mathbf{t}_i) X^{q^{i+1}} \right).
\end{aligned}$$

---

[2] Note, that the Steenrod algebra acts on the additional variable X as though it had degree 1:

$$\mathcal{P}^i(X) = \begin{cases} X & \text{if } i = 0 \\ X^q & \text{if } i = 1 \\ 0 & \text{otherwise} \end{cases}.$$

The preceding proposition tells us that $\Delta(X) \mid \mathcal{P}^{\alpha}\Delta(X)$ if and only if

$$\sum_{i=0}^{m} \mathcal{P}^{\alpha}(\mathbf{t}_i)\mathcal{P}^{\Delta_i} + \sum_{i=0}^{m} \mathcal{P}^{\alpha-q^i}(\mathbf{t}_i)\mathcal{P}^{\Delta_{i+1}}$$

vanishes on H^*. But notice: the left hand side in this equation is just good old $\Delta(\alpha, \mathbf{t}_0, \ldots, \mathbf{t}_{m-1}, 1)$, which is zero by Proposition 1.2.4. •

We need the following generalization of Lemma 5.8 in [1].

LEMMA 5.1.6 : *Let $\mathbb{K}^*$ be a graded field over the Steenrod algebra, and $f(X) \in \mathbb{K}^*[X]$ a 1-graded polynomial of degree m in X with roots $\lambda_1, \ldots, \lambda_m$ of degree 1. If, for any $\alpha \geq 0$,*

$$f(X) \mid \mathcal{P}^{\alpha} f(X),$$

then the roots are unstable elements, i.e., fulfil the unstability condition.

PROOF: Write the polynomial $f(X)$ in factored form

$$f(X) = k(X - \lambda_1) \cdots (X - \lambda_m),$$

where $k \in \mathbb{K}^*$. Since $f(X) \mid \mathcal{P}^{\alpha} f(X)$ for any α we have

$$f(X) \mid P(\xi)f(X) = \left(P(\xi)(k)\right)\left(P(\xi)(X - \lambda_1)\right) \cdots \left(P(\xi)(X - \lambda_m)\right).$$

Hence, for every $i = 1, \ldots, m$, there exists an $j \in \{1, \ldots, m\}$, such that

$$P(\xi)(X - \lambda_j)\Big|_{X=\lambda_i} = 0.$$

From

$$\mathcal{P}^{\alpha}(X - \lambda_j) = \mathcal{P}^{\alpha}(X) - \mathcal{P}^{\alpha}(\lambda_j) = \begin{cases} X - \lambda_j & \text{for } \alpha = 0 \\ X^q - \mathcal{P}^1(\lambda_j) & \text{for } \alpha = 1 \\ -\mathcal{P}^{\alpha}(\lambda_j) & \text{for } \alpha > 1, \end{cases}$$

we obtain by setting $X = \lambda_i$

$$\lambda_i = \lambda_j$$
$$\lambda_i^q = \mathcal{P}^1(\lambda_j)$$
$$\mathcal{P}^{\alpha}(\lambda_j) = 0 \qquad \text{for } \alpha > 1.$$

This shows that the roots are unstable. •

We come back to our problem of trying to show that the polynomial algebra

$$\mathbb{F}[\mathbf{t}_0, \ldots, \mathbf{t}_{n-1}]$$

is isomorphic to the Dickson algebra $\mathcal{D}^*(n)$ *as algebra over the Steenrod algebra*. To continue our investigation of the Steenrod powers $\mathcal{P}^i(\mathbf{t}_j)$, recall the Δ-polynomial

$$\Delta(X) = \mathbf{t}_0 X + \mathbf{t}_1 X^q + \cdots + \mathbf{t}_m X^{q^m}.$$

By Lemma 5.1.5 we have for any $\alpha \in \mathbb{N}_0$

$$\Delta(X) \mid \mathcal{P}^{\alpha}\Delta(X),$$

hence, by Lemma 5.1.6, the polynomial $\Delta(X)$ has unstable roots. Since $\Delta(X)$ is monic, the coefficients

$$\mathbf{t}_0, \ldots, \mathbf{t}_{m-1}, \mathbf{t}_m = 1$$

are polynomials in these roots, so are also unstable elements in $FF(\mathrm{H}^*)$. We apply Proposition 1.2.4 and get

$$\begin{aligned}&\Delta(\alpha, \mathbf{t}_0, \ldots, \mathbf{t}_{m-1}, \mathbf{t}_m = 1)\\ &\quad = \mathcal{P}^{\alpha}(\mathbf{t}_0)\mathcal{P}^{\Delta_0} + \cdots + \mathcal{P}^{\alpha}(\mathbf{t}_m)\mathcal{P}^{\Delta_m} + \sum_{i=0}^{m} \mathcal{P}^{\alpha - q^i}(\mathbf{t}_i)\mathcal{P}^{\Delta_{i+1}} = 0.\end{aligned}$$

Hence, by minimality of m and Observation 5.1.1, we have for the $(m+1)$-tuple of coefficients of $\Delta_{\alpha} = \Delta(\alpha, \mathbf{t}_0, \ldots, \mathbf{t}_m)$,

$$\begin{aligned}&\Big(\mathcal{P}^{\alpha}(\mathbf{t}_0),\ \mathcal{P}^{\alpha}(\mathbf{t}_1) + \mathcal{P}^{\alpha-1}(\mathbf{t}_0),\ \mathcal{P}^{\alpha}(\mathbf{t}_2) + \mathcal{P}^{\alpha-q}(\mathbf{t}_1), \ldots\\ &\qquad \ldots, \mathcal{P}^{\alpha}(\mathbf{t}_{j+1}) + \mathcal{P}^{\alpha-q^j}(\mathbf{t}_j), \ldots, \mathcal{P}^{\alpha}(\mathbf{t}_m) + \mathcal{P}^{\alpha-q^{m-1}}(\mathbf{t}_{m-1})\Big)\\ &= \begin{cases} \mathcal{P}^{\alpha}(\mathbf{t}_m) \cdot (\mathbf{t}_0, \ldots, \mathbf{t}_{m-1}, 1) & \text{for } \alpha < q^{m-1}\\ \Big(\mathcal{P}^{\alpha}(\mathbf{t}_m) + \mathcal{P}^{\alpha-q^{m-1}}(\mathbf{t}_{m-1})\Big) \cdot (\mathbf{t}_0, \ldots, \mathbf{t}_{m-1}, 1) & \text{for } q^{m-1} \le \alpha < q^m\\ 0 \cdot (\mathbf{t}_0, \ldots, \mathbf{t}_{m-1}, 1) & \text{otherwise,}\end{cases}\end{aligned}$$

where the last statement is true for degree reasons.

Let's collect this in a proposition:

PROPOSITION 5.1.7: *With the preceding notations we have*

$$\begin{aligned}&\Big(\mathcal{P}^{\alpha}(\mathbf{t}_0),\ \mathcal{P}^{\alpha}(\mathbf{t}_1) + \mathcal{P}^{\alpha-1}(\mathbf{t}_0),\ \mathcal{P}^{\alpha}(\mathbf{t}_2) + \mathcal{P}^{\alpha-q}(\mathbf{t}_1), \ldots\\ &\qquad \ldots, \mathcal{P}^{\alpha}(\mathbf{t}_{j+1}) + \mathcal{P}^{\alpha-q^j}(\mathbf{t}_j), \ldots, \mathcal{P}^{\alpha}(\mathbf{t}_m) + \mathcal{P}^{\alpha-q^{m-1}}(\mathbf{t}_{m-1})\Big)\\ &= \begin{cases} (\mathbf{t}_0, \ldots, \mathbf{t}_{m-1}, 1) & \text{for } \alpha = 0\\ \Big(\mathcal{P}^{\alpha-q^{m-1}}(\mathbf{t}_{m-1})\Big) \cdot (\mathbf{t}_0, \ldots, \mathbf{t}_{m-1}, 1) & \text{for } q^{m-1} \le \alpha < q^m\\ 0 \cdot (\mathbf{t}_0, \ldots, \mathbf{t}_{m-1}, 1) & \text{otherwise,}\end{cases}\end{aligned}$$

from which we may read off recursively the Steenrod powers of the elements $\mathbf{t}_0, \ldots, \mathbf{t}_m$.

PROOF: Since $\mathbf{t}_m = 1$,

$$\mathcal{P}^{\alpha}(\mathbf{t}_m) = \begin{cases} \mathbf{t}_m & \text{if } \alpha = 0\\ 0 & \text{otherwise}\end{cases}$$

and the formulae can be read off those we derived above. •

We can now prove that $\mathbb{F}[\mathbf{t}_0, \ldots, \mathbf{t}_{m-1}]$ is a Dickson algebra *in the category of unstable algebras over* $\mathcal{P}^*$, and hence establishes the first part of the main goal of this section.

THEOREM 5.1.8: *The isomorphism of $\mathbb{F}$-algebras given by*

$$\begin{array}{rcccl} \varphi : \mathbb{F}_q[\mathbf{t}_0, \dots, \mathbf{t}_{m-1}] & \longrightarrow & \mathcal{D}^*(m) = \mathbb{F}_q[\mathbf{d}_{m,0}, \dots, \mathbf{d}_{m,m-1}] \\ \mathbf{t}_i & \longmapsto & (-1)^{m-i+1}\mathbf{d}_{m,i} \end{array}$$

is an isomorphism of algebras over the Steenrod algebra.

PROOF: Is is clear that φ is an isomorphism of $\mathbb{F}$-algebras. What needs to be shown is that φ commutes with the action of the Steenrod algebra, i.e.,

$$\varphi \mathcal{P}^\alpha(\mathbf{t}_j) = \mathcal{P}^\alpha \varphi(\mathbf{t}_j)$$

for every choice of α or j. We prove this by induction on α. If $\alpha = 0$ there is nothing to show, since $\mathcal{P}^0$ is the identity operator. So, we can assume that $\alpha \geq 1$. If $\alpha \geq q^m$ both sides are zero for degree reasons. So, assume that $1 \leq \alpha < q^m$.

Again, we will make use of the recursion formulae for the Steenrod powers of the $\mathbf{t}_0, \dots, \mathbf{t}_m$ given by the $\Delta(\alpha, \mathbf{t}_0, \dots, \mathbf{t}_m)$-relation from Proposition 1.2.4, resp. Proposition 5.1.7. If $1 \leq q^j \leq \alpha < q^{j+1} \leq q^{m-1}$, then

$$\Big(\mathcal{P}^\alpha(\mathbf{t}_0),\ \mathcal{P}^\alpha(\mathbf{t}_1) + \mathcal{P}^{\alpha-1}(\mathbf{t}_0),\ \mathcal{P}^\alpha(\mathbf{t}_2) + \mathcal{P}^{\alpha-q}(\mathbf{t}_1), \dots \\ \dots,\ \mathcal{P}^\alpha(\mathbf{t}_{j+1}) + \mathcal{P}^{\alpha-q^j}(\mathbf{t}_j),\ \mathcal{P}^\alpha(\mathbf{t}_{j+2}), \dots,\ \mathcal{P}^\alpha(\mathbf{t}_{m-1}),\ 0\Big) = 0.$$

Therefore, if $1 \leq q^j \leq \alpha < q^{j+1} \leq q^{m-1}$, then

$$\mathcal{P}^\alpha(\mathbf{t}_i) = \begin{cases} 0 & \text{for } i = 0,\ j+2, \dots, m \\ -\mathcal{P}^{\alpha-q^{i-1}}(\mathbf{t}_{i-1}) & i = 1, \dots, j+1. \end{cases}$$

Hence, by induction

$$\begin{aligned} \varphi(\mathcal{P}^\alpha(\mathbf{t}_i)) &= \begin{cases} \varphi(0) & \text{for } i = 0,\ j+2, \dots, m \\ -\varphi(\mathcal{P}^{\alpha-q^{i-1}}(\mathbf{t}_{i-1})) & i = 1, \dots, j+1 \end{cases} \\ &= \begin{cases} 0 & \text{for } i = 0,\ j+2, \dots, m \\ -(-1)^{m-i}\mathcal{P}^{\alpha-q^{i-1}}(\mathbf{d}_{m,i-1}) & i = 1, \dots, j+1 \end{cases} \\ &= \begin{cases} (-1)^{m-i+1}\mathcal{P}^\alpha(\mathbf{d}_{m,i}) & \text{for } i = 0,\ j+2, \dots, m \\ -(-1)^{m-i}\mathcal{P}^{\alpha-q^{i-1}}(\mathbf{d}_{m,i-1}) & i = 1, \dots, j+1 \end{cases} \\ &= \begin{cases} (-1)^{m-i+1}\mathcal{P}^\alpha(\mathbf{d}_{m,i}) & \text{for } i = 0,\ j+2, \dots, m \\ (-1)^{m-i+1}\mathcal{P}^\alpha(\mathbf{d}_{m,i}) & i = 1, \dots, j+1 \end{cases} \\ &= \mathcal{P}^\alpha(\varphi(\mathbf{t}_i)). \end{aligned}$$

where we made use of the explicit formulae of the action of the Steenrod algebra on the Dickson classes given in Proposition A.2.1 of the appendix.

Finally, lets examine the case where $q^{m-1} \le \alpha < q^m$. Then, by Proposition 5.1.7, the above equation reads as follows

$$\Big(\mathcal{P}^{\alpha}(\mathbf{t}_0),\ \mathcal{P}^{\alpha}(\mathbf{t}_1) + \mathcal{P}^{\alpha-1}(\mathbf{t}_0),\ \mathcal{P}^{\alpha}(\mathbf{t}_2) + \mathcal{P}^{\alpha-q}(\mathbf{t}_1), \ldots$$
$$\ldots,\ \mathcal{P}^{\alpha}(\mathbf{t}_{m-1}) + \mathcal{P}^{\alpha-q^{m-2}}(\mathbf{t}_{m-2}),\ \mathcal{P}^{\alpha-q^{m-1}}(\mathbf{t}_{m-1})\Big)$$
$$= \mathcal{P}^{\alpha-q^{m-1}}(\mathbf{t}_{m-1})(\mathbf{t}_0, \ldots, \mathbf{t}_{m-1}, 1).$$

So we get by induction

$$\begin{aligned}
&\varphi(\mathcal{P}^{\alpha}(\mathbf{t}_j)) \\
&= \begin{cases} \varphi\left(\mathcal{P}^{\alpha-q^{m-1}}(\mathbf{t}_{m-1})\mathbf{t}_0\right) & \text{for } j = 0 \\ \varphi\left(\mathcal{P}^{\alpha-q^{m-1}}(\mathbf{t}_{m-1})\mathbf{t}_j - \mathcal{P}^{\alpha-q^{j-1}}(\mathbf{t}_{j-1})\right) & \text{otherwise} \end{cases} \\
&= \begin{cases} \left(\mathcal{P}^{\alpha-q^{m-1}}(\varphi(\mathbf{t}_{m-1}))\right)\varphi(\mathbf{t}_0) & \text{for } j = 0 \\ \left(\mathcal{P}^{\alpha-q^{m-1}}(\varphi(\mathbf{t}_{m-1}))\right)\varphi(\mathbf{t}_j) - \mathcal{P}^{\alpha-q^{j-1}}(\varphi(\mathbf{t}_{j-1})) & \text{otherwise} \end{cases} \\
&= \begin{cases} (-1)^{m+1}\mathcal{P}^{\alpha-q^{m-1}}(\mathbf{d}_{m,m-1})\mathbf{d}_{m,0} & \text{for } j = 0 \\ (-1)^{m-j+1}\mathcal{P}^{\alpha-q^{m-1}}(\mathbf{d}_{m,m-1})\mathbf{d}_{m,j} - (-1)^{m-j+1}\mathcal{P}^{\alpha-q^{j-1}}(\mathbf{d}_{m,j-1}) & \text{otherwise} \end{cases} \\
&= \begin{cases} (-1)^{m+1}\mathcal{P}^{\alpha}(\mathbf{d}_{m,0}) & \text{for } j = 0 \\ (-1)^{m-j+1}\mathcal{P}^{\alpha}(\mathbf{d}_{m,j}) & \text{otherwise} \end{cases} \\
&= \mathcal{P}^{\alpha}(\varphi(\mathbf{t}_j)),
\end{aligned}$$

where we again made use of Proposition A.2.1 in the appendix. This proves the theorem. •

We summarize the results of this section in the following theorem.

THEOREM 5.1.9: *Let* H^* *be a* Δ*-finite, unstable integral domain with* Δ*-relation*

$$h_0\mathcal{P}^{\Delta_0} + \cdots + h_m\mathcal{P}^{\Delta_m} = 0.$$

Then there exists a unique Dickson algebra $\mathcal{D}^*(m)$ *of Krull dimension* m *in the field of fractions of* H^* *generated by* $\mathbf{d}_{m,i} = (-1)^{m-i+1}\frac{h_i}{h_m} \in FF(\mathrm{H}^*)$, $i = 0, \ldots, m-1$. •

REMARK: We could have proved the preceding theorem also by simply noting that

$$(-1)^m\mathbf{d}_{m,0}\mathcal{P}^{\Delta_0} + \cdots + (-1)\mathbf{d}_{m,m-1}\mathcal{P}^{\Delta_{m-1}} + \mathcal{P}^{\Delta_m} = 0$$

is the Δ-relation in $\mathcal{D}^*(m)$, see the remark after Theorem 1.2.3. Hence the analogous formulae as in Proposition 5.1.7 hold for the Dickson classes (with appropriate signs) as well for $\mathbf{t}_0, \ldots, \mathbf{t}_{m-1}$.

REMARK: The proof also shows that the Dickson algebra of Krull dimension m we have just found inside $FF(\mathrm{H}^*)$ is maximal. For otherwise there would be a $k > m$ such that

$$\varphi : \mathcal{D}^*(k) \hookrightarrow FF(\mathrm{H}^*)$$

is a monomorphism. But then

$$(-1)^k \varphi(\mathbf{d}_{k,0})\mathcal{P}^{\Delta_0} + \cdots + (-1)\varphi(\mathbf{d}_{k,k-1})\mathcal{P}^{\Delta_{k-1}} + \mathcal{P}^{\Delta_k} = 0$$

would also be a linear relation of *minimal* length, which is a contradiction. This means that $m = m(\mathrm{H}^*)$ is an invariant of H^*.

REMARK: On the other hand, one could use the above theorem to again prove that the Δ-relation for the Dickson algebra $\mathcal{D}^*(m)$ reads as follows:

$$(-1)^m \mathbf{d}_{m,0}\mathcal{P}^{\Delta_0} + \cdots + (-1)\mathbf{d}_{m,m-1}\mathcal{P}^{\Delta_{m-1}} + \mathcal{P}^{\Delta_m} = 0.$$

This is because, up to a sign, it is the image of the Δ-relation in $FF(\mathrm{H}^*)$ under the map given in the theorem.

5.2 Preparing the Embedding Theorem après Smith-Switzer

In this section we prepare for the proof of the general version of the Embedding Theorem following the path described by [40]. We are going to establish the proposition in their Section 2 (see also [36] Theorem 10.5.5) for arbitrary Galois fields.

Consider a Δ-relation in H^*

$$h_0\mathcal{P}^{\Delta_0} + \cdots + h_m\mathcal{P}^{\Delta_m} = 0,$$

and recall from Section 5.1 the Δ-polynomial

$$\Delta(X) := \mathbf{t}_0X + \mathbf{t}_1X^q + \cdots + \mathbf{t}_mX^{q^m} \in FF(\mathrm{H}^*)[X],$$

where $\mathbf{t}_i = \frac{h_i}{h_m}$ for $i = 0, \ldots, m$, and $\mathbf{t}_m = 1$. It has a splitting field $\mathbb{E}^*$, and, since the formal derivative

$$\frac{\mathrm{d}}{\mathrm{d}X}\Delta(X) = \mathbf{t}_0 \neq 0$$

is not zero, the field extension $\mathbb{E}^*/FF(\mathrm{H}^*)$ is separable. Hence the Steenrod algebra action on H^* can be extended uniquely to $\mathbb{E}^*$ by the Separable Extension Lemma, Lemma 2.2.2. Consider the subset of zeros of this polynomial

$$V := \{v \in \mathbb{E}^* \mid \Delta(v) = 0\} \subseteq \mathbb{E}^*.$$

Since $\Delta(X)$ has no multiple roots, V contains q^m elements. Since $\Delta(X)$ is a q-polynomial we have that

$$V = \ker\{\Delta(X) : \mathbb{E}^* \longrightarrow \mathbb{E}^*\}$$

is an $\mathbb{F}_q$-vector subspace of $\mathbb{E}^*$ of dimension m over $\mathbb{F}_q$. Hence, since $\mathbf{t}_m = 1$,

$$\Delta(X) = X \prod_{v \in V \setminus \{0\}} (X - v).$$

Let $z_1, \ldots, z_m$ be a basis of V as a vector space over $\mathbb{F}_q$.

We are now fit to prove the following theorem, compare Theorem 10.5.5 in [36]:

THEOREM 5.2.1 : *Let* H^* *be a* Δ*-finite unstable integral domain, with* Δ*-polynomial*

$$\Delta(X) = \mathbf{t}_0 X + \mathbf{t}_1 X^q + \cdots + \mathbf{t}_m X^{q^m}.$$

Let $\mathbb{E}^*$ *be a splitting field for* $\Delta(X)$ *over* $FF(\mathrm{H}^*)$*, and let* $z_1, \ldots, z_m$ *be an* $\mathbb{F}_q$*-basis of the vector subspace* V *of* $\mathbb{E}^*$ *of roots of* $\Delta(X)$*. Then*

(1) $z_1, \ldots, z_m$ *are unstable elements, hence* $\mathbb{F}[z_1, \ldots, z_m]$ *is an unstable algebra over the Steenrod algebra.*
(2) $z_1, \ldots, z_m$ *are algebraically independent.*
(3) $\mathcal{D}^* = \mathbb{F}[z_1, \ldots, z_m]^{\mathrm{GL}(m,\, \mathbb{F})} \hookrightarrow \mathbb{F}[z_1, \ldots, z_m]$ *is an inclusion of unstable algebras over the Steenrod algebra.*
(4) *If* H^* *is in addition Noetherian, then* $\mathbb{F}[z_1, \ldots, z_m]$ *is integral over* H^**, i.e., the elements* $z_1, \ldots, z_m$ *are roots of a monic polynomial with coefficients in* H^**.*

PROOF: We take each point in turn.

AD (1) : By Lemma 5.1.5 for any $\alpha \geq 0$ we have

$$\Delta(X) \mid \mathcal{P}^{\alpha} \Delta(X),$$

which in turn implies, by Lemma 5.1.6, that $\Delta(X)$ has unstable roots. In particular it follows that

$$\mathcal{P}^{\alpha}(z_j) = \begin{cases} z_j & \text{for } \alpha = 0 \\ z_j^q & \text{for } \alpha = 1 \\ 0 & \text{otherwise} \end{cases}$$

for $j = 1, \ldots, m$.

AD (2) : Taken together with what we just proved, the determinant of the generalized Jacobian matrix (compare Theorem A.4.1) is

$$\det \left(\mathcal{P}^{\Delta_i}(z_j)\right)_{i=0,\ldots,m-1}^{j=1,\ldots,m} = \det \begin{bmatrix} z_1 & \cdots & z_m \\ \vdots & \ddots & \vdots \\ z_1^{q^{m-1}} & \cdots & z_m^{q^{m-1}} \end{bmatrix} \neq 0.$$

Therefore $z_1, \ldots, z_m$ are algebraically independent.[3]

[3] Recall that this determinant was calculated by L. E. Dickson in [9] and is the Euler class $\mathbf{E}_n$, for further information see Section 8.1 in [36].

AD (3) : By part (1) and (2) we have that

$$\mathbb{F}[z_1, \ldots, z_m]$$

is an unstable algebra over the Steenrod algebra. Therefore

$$\mathbb{F}[z_1, \ldots, z_m]^{\mathrm{GL}(m,\, \mathbb{F})} \subseteq \mathbb{F}[z_1, \ldots, z_m]$$

is an inclusion as algebras over the Steenrod algebra for the tautological action of $\mathrm{GL}(m, \mathbb{F})$.

AD (4) : Recall, from the proof of Theorem 1.2.1, that there exists an integer g such that for any $h \in \mathrm{H}^*$

$$\mathcal{P}^{\Delta_{g+1}}(h) = a_0\mathcal{P}^{\Delta_0}(h) + \cdots + a_g\mathcal{P}^{\Delta_g}(h)$$

for certain coefficients $a_0, \ldots, a_g \in \mathrm{H}^*$. By Proposition 2.2.2 this holds for $h \in \mathbb{E}^*$ and, in particular, for any $h \in V$. Hence we have

$$z_j^{q^{g+1}} = a_0 z_j + \cdots + a_g z_j^{q^g}$$

for $j = 1, \ldots, m$, i.e., $\mathbb{F}[z_1, \ldots, z_m]$ is integral over H^*. •

5.3 The Embedding Theorem

From what we have done so far we have the following diagram of unstable algebras over the Steenrod algebra:

$$\begin{array}{ccc} \mathcal{D}^*(m) & & \mathrm{H}^* \\ \uparrow \text{integral} & & \uparrow \\ \mathbb{F}[z_1, \ldots, z_m] & \underset{\psi}{\hookrightarrow} & \mathrm{H}^*\langle z_1, \ldots, z_m\rangle =: \mathrm{A}^*. \end{array}$$

This notation will remain fixed throughout this section.

For the key results of this section we need the following lemma.

LEMMA 5.3.1: H^* *is* $\mathcal{P}^*$*-inseparably closed if and only if* A^* *is.*

PROOF: Let H^* be $\mathcal{P}^*$-inseparably closed, and, suppose to the contrary that A^* were not $\mathcal{P}^*$-inseparably closed. Then looking at the fields of fractions we would get

$$\begin{array}{ccc} FF(\mathrm{H}^*) & \underset{\text{separable}}{\hookrightarrow} & FF(A^*) \\ & & \downarrow {\scriptstyle \mathcal{P}^*\text{-purely insep.}} \\ & & FF(\sqrt[\mathcal{P}^*]{A^*}), \end{array}$$

where the upper map is $\mathcal{P}^*$-separable, because $FF(A^*) = \mathbb{E}^*$ is the splitting field of the separable polynomial $\Delta(X)$, and the pure $\mathcal{P}^*$-inseparability of the map downwards follows from Lemma 4.2.5. In such a case we could find an intermediate field $\mathbb{K}^*$ between $FF(\mathrm{H}^*)$ and $FF(\sqrt[\mathcal{P}^*]{A^*})$, such that

$$FF(\mathrm{H}^*) \subseteq \mathbb{K}^*$$

is $\mathcal{P}^*$-purely inseparable, while

$$\mathbb{K}^* \subseteq FF(\sqrt[\mathcal{P}^*]{A^*})$$

is $\mathcal{P}^*$-separable, see the remarks at the end of Section 2.3. However, by Corollary 4.2.7, $FF(\mathrm{H}^*)$ is $\mathcal{P}^*$-inseparably closed, which means that $\mathbb{K}^* = FF(\mathrm{H}^*)$ and

$$FF(\mathrm{H}^*) \hookrightarrow FF(\sqrt[\mathcal{P}^*]{A^*})$$

is $\mathcal{P}^*$-separable. This is a contradiction.

To prove the converse assume that H^* is not $\mathcal{P}^*$-inseparably closed. Suppose A^* were $\mathcal{P}^*$-inseparably closed. Then, taken together, Lemma 4.2.5 and Corollary 4.2.7 imply a commutative diagram

$$\begin{array}{ccc} FF(\mathrm{H}^*) & \underset{\mathcal{P}^*\text{-purely insep.}}{\hookrightarrow} & FF(\sqrt[\mathcal{P}^*]{\mathrm{H}^*}) \\ \downarrow & & \downarrow \\ FF(A^*) & = & FF(\sqrt[\mathcal{P}^*]{A^*}). \end{array}$$

Therefore the field extension $FF(A^*)/FF(\mathrm{H}^*)$ must be $\mathcal{P}^*$-inseparable. This is a contradiction, since $FF(A^*) = \mathbb{E}^*$ is a splitting field of a $\mathcal{P}^*$-separable polynomial, namely of $\Delta(X)$. •

LEMMA 5.3.2: *If ψ is integral, then A^* is $\mathcal{P}^*$-inseparably closed.*

PROOF: If

$$\psi : \mathbb{F}[z_1, \dots, z_m] \hookrightarrow A^*$$

is an integral extension, then the Integral Closure Theorem 3.2.2 tells us that ψ is an isomorphism, because A^* is an unstable integral domain. Hence A^* is $\mathcal{P}^*$-inseparably closed because, by Corollary 4.2.8, the polynomial ring $\mathbb{F}[z_1, \dots, z_m]$ is. •

The following theorem is the key to the Embedding Theorem and the Little Imbedding Theorem in Section 7.4.

THEOREM 5.3.3: *With the same notation as above, if A^* is $\mathcal{P}^*$-inseparably closed, then ψ is an isomorphism.*

PROOF: Let A^* be $\mathcal{P}^*$-inseparably closed. By construction of A^*, both algebras have the same Δ-relation of finite length m. Therefore, at the level of fields of fractions, we have

$$\dim\left(\Delta(\mathbb{F}(z_1, \dots, z_m))\right) = m = \dim\left(\Delta(FF(A^*))\right)$$

By Corollary 4.2.7, the field of fraction $FF(A^*)$ of A^* is $\mathcal{P}^*$-inseparably closed, because A^* is. Hence Corollary 2.4.2 tells us that the two fields must be equal

$$\mathbb{F}(z_1, \dots, z_m) = FF(A^*).$$

Therefore, by Lemma 3.1.3,

$$A^* \subseteq \mathcal{U}n\left(FF(A^*)\right) = \mathcal{U}n\left(\mathbb{F}(z_1, \dots, z_m)\right) = \mathbb{F}[z_1, \dots, z_m] \overset{\psi}{\hookrightarrow} A^*$$

Hence both algebras are equal and ψ is the identity map even. •

As a corollary from this we get the Embedding Theorem over arbitrary Galois fields, see Theorem 1.1 in [1] for the case of algebras over prime fields.

THEOREM 5.3.4 (Embedding Theorem, Version I): *Let* H^* *be an* $\mathcal{P}^*$*-inseparably closed unstable integral domain of* Δ*-length* m. *Then* H^* *can be embedded into a polynomial ring with* m *linear generators. Moreover, the embedding is integral if and only if* H^* *is Noetherian.*

PROOF: The algebra A^* is $\mathcal{P}^*$-inseparably closed by Lemma 5.3.1, because H^* is $\mathcal{P}^*$-inseparably closed. So, Theorem 5.3.3 shows that the map ψ is an isomorphism, i.e.,

$$\mathrm{H}^* \hookrightarrow A^* \overset{\psi}{\cong} \mathbb{F}[z_1, \ldots, z_m]$$

is the embedding we wanted.

If H^* is Noetherian then the embedding $\mathrm{H}^* \hookrightarrow A^*$ is integral by Theorem 5.2.1 (4). Conversely, assume that H^* is not Noetherian. The polynomial ring $\mathbb{F}[z_1, \ldots, z_m]$ is finitely generated as an algebra over $\mathbb{F}$ and, a fortiori, over H^*. So, if the embedding $\mathrm{H}^* \hookrightarrow \mathbb{F}[z_1, \ldots, z_m]$ were integral, it were also finite, i.e., $\mathbb{F}[z_1, \ldots, z_m]$ were a finite H^*-module. This would imply that H^* were Noetherian, by Theorem 2 of [15]. This is a contradiction. •

REMARK: If H^* and A^* are Noetherian, they are finitely generated $\mathbb{F}$-algebras. Hence the integral embedding we have constructed is finite (as a map of H^*-modules). (Just that you don't miss the point.) Moreover, a posteriori we see that the Δ-length m of H^* is equal to its Krull dimension m.

For later use we collect the technical results of this section in a theorem.

THEOREM 5.3.5: *Let* H^* *be a* Δ*-finite unstable integral domain. Let* A^* *be obtained from* H^* *by adjoining all roots* $z_1, \ldots, z_m$ *of the* Δ*-polynomial of* H^*. *The following statements are equivalent:*

(1) H^* *is* $\mathcal{P}^*$*-inseparably closed.*
(2) A^* *is* $\mathcal{P}^*$*-inseparably closed.*
(3) *The inclusion* $\psi : \mathbb{F}[z_1, \ldots, z_m] \hookrightarrow A^*$ *is an isomorphism.*

Moreover, if H^* *is Noetherian, the above statements imply*[4] *that the* Δ*-length* m *of* H^* *is equal to its Krull dimension.*

PROOF: The equivalence of the first two statements is the contents of Lemma 5.3.1. Statement (3) implies (2) by Lemma 5.3.2. In Theorem 5.3.3 we have seen that (2) ⇒ (3). Finally if the inclusion ψ is an isomorphism and H^* Noetherian, then

$$m = \dim(\mathbb{F}[z_1, \ldots, z_m]) = \dim(A^*) = \dim(\mathrm{H}^*),$$

where the last equation follows from part (4) of Theorem 5.2.1. •

[4] We will see later that also the converse is true, see Corollary 6.1.4.

In Theorem 5.3.4 we could drop the condition on H^* of being $\mathcal{P}^*$-inseparably closed. We would consider the embedding

$$H^* \underset{\text{integral}}{\hookrightarrow} \sqrt[\mathcal{P}^*]{H^*},$$

and since $\sqrt[\mathcal{P}^*]{H^*}$ is $\mathcal{P}^*$-inseparably closed ..... Whoops! There is a problem. Why is $\sqrt[\mathcal{P}^*]{H^*}$ Noetherian? Or, at least Δ-finite?

Chapter 6
Noetherianess, the Embedding Theorem II and Turkish Delights

We closed the preceding chapter with a proof of the Embedding Theorem, Theorem 5.3.4. However, we had to assume that the algebra in question, H^*, was $\mathcal{P}^*$-inseparably closed. Since we can embed any H^* in its inseparable closure $\sqrt[\mathcal{P}^*]{H^*}$ one might be tempted to say, well then embed the inseparable closed $\sqrt[\mathcal{P}^*]{H^*}$ into $\mathbb{F}[V]$ and we are done. But wait! We can apply the Embedding Theorem 5.3.4 (in this version I) only if $\sqrt[\mathcal{P}^*]{H^*}$ is Noetherian. We don't know yet when this happens.[1] So, this is one of the main topics of this chapter. This allows us to prove the Embedding Theorem in its full generality. Moreover, we will be rewarded with some gorgeous results about the $\mathcal{P}^*$-invariant prime spectrum, which was part of the original motivation for these investigations.

Throughout the whole chapter H^* is an unstable algebra over the Steenrod algebra.

6.1 Noetherianess

The main goal of this section is to prove that an unstable integral domain H^* is Noetherian if and only if its $\mathcal{P}^*$-inseparable closure $\sqrt[\mathcal{P}^*]{H^*}$ is Noetherian. This makes then the proof of the Embedding Theorem in its full generality easy.

Recall from Chapter 4 that we have a chain of unstable algebras

$$H^* = H_0^* \overset{\varphi_0}{\hookrightarrow} H_1^* \overset{\varphi_1}{\hookrightarrow} \cdots \overset{\varphi_{i-1}}{\hookrightarrow} H_i^* \overset{\varphi_i}{\hookrightarrow} \cdots,$$

where each extension is integral, so that

$$\varphi : H^* \hookrightarrow \sqrt[\mathcal{P}^*]{H^*} := \operatorname{colim}(H_i^*)$$

[1] This is precisely the reason why we emphasized up to now when the algebras involved had to be Noetherian, or when a weaker condition would be sufficient. That made the writing sometimes a bit clumsy and the reading a bit awkward, I know. I hope you forgive me now!

is also an integral ring extension.

In Proposition 4.2.3 we have seen that if $\sqrt[\mathcal{P}^*]{\mathrm{H}^*}$ is Noetherian then so is H^*. The converse of this proposition is a lot more difficult. Let's assume that H^* is the least pretentious graded ring you can think of, i.e., a polynomial ring in one linear generator over a Galois field $\mathbb{F}$. Still, this would not imply (from classical tic-tac-toe or such) that $\sqrt[\mathcal{P}^*]{\mathrm{H}^*}$ is Noetherian (just because the extension is integral). This is illustrated by the following gorgeous example of Leslie G. Roberts, [31].

EXAMPLE 1 (Leslie G. Roberts): Let $\mathbb{F}[x]$ be a polynomial ring in one indeterminate of degree 1 over the field $\mathbb{F}$. Then we map this ring diagonally into an infinite product of polynomial rings, $\mathbb{F}[x_i]$, in one generator of degree 1 over the field $\mathbb{F}$

$$\begin{array}{rcl} \Delta : \mathbb{F}[x] & \hookrightarrow & \prod_{i\in\mathbb{N}} \mathbb{F}[x_i] \\ x & \longmapsto & (x_1, x_2, \ldots) \end{array}$$

Denote by X_i the element of $\prod_{i\in\mathbb{N}} \mathbb{F}[x_i]$ that is x_i in the i-th coordinate and zero elsewhere. Then we take the $\mathbb{F}$-subalgebra B in $\prod_{i\in\mathbb{N}} \mathbb{F}[x_i]$ generated by these X_i's and the image of x, $\Delta(x)$,

$$B := \mathbb{F}\langle \Delta(x), X_1, X_2, \ldots \rangle.$$

Then B is certainly not Noetherian, because the ideal generated by the elements of positive degree is not finitely generated. However, B is integral over $\Delta(\mathbb{F}[x]) \cong \mathbb{F}[x]$, because

$$X_i^2 - \Delta(x)X_i = 0 \quad \forall\, i \in \mathbb{N}.$$

Note that we could even put a $\mathcal{P}^*$-action onto the algebras involved; that wouldn't make any difference.

We start with an investigation of the Δ-length of a filtered colimit.

THEOREM 6.1.1: *Let*

$$\mathrm{H}_0^* \subseteq \mathrm{H}_1^* \subseteq \mathrm{H}_2^* \subseteq \cdots$$

be an increasing chain of integral domains. If all H_i^* are Δ-finite of the same length m, then their direct limit $\mathrm{colim}(\mathrm{H}_i^*)$ is Δ-finite of the same length m. Moreover we can choose a Δ-relation with coefficients in H_0^*, which is valid for all H_i^* and for the colimit.

PROOF: Denote by

$$h_0(i)\mathcal{P}^{\Delta_0} + h_1(i)\mathcal{P}^{\Delta_1} + \cdots + h_m(i)\mathcal{P}^{\Delta_m} = 0$$

a Δ-relation for H_i^*, $i = 0, 1, \ldots$. Considered as equations in the respective field of fractions, $FF(\mathrm{H}_i^*)$, we can divide by the respective leading coefficient $h_m(i)$

to get *uniquely determined* relations

$$\frac{h_0(i)}{h_m(i)}\mathcal{P}^{\Delta_0} + \frac{h_1(i)}{h_m(i)}\mathcal{P}^{\Delta_1} + \cdots + \frac{h_{m-1}(i)}{h_m(i)}\mathcal{P}^{\Delta_{m-1}} + \mathcal{P}^{\Delta_m} = 0$$

for every i. Since a Δ-relation for H_i^* certainly evaluates to zero also on the subalgebra H_{i-1}^*, we conclude that all these must coincide:

$$\frac{h_j(i)}{h_m(i)} = \frac{h_j(i')}{h_m(i')} \quad \forall\, i,\ i',\ j,$$

(compare the remark after the Δ-Theorem 1.2.1). Therefore

$$h_m(0)\left(\frac{h_0(0)}{h_m(0)}\mathcal{P}^{\Delta_0} + \frac{h_1(0)}{h_m(0)}\mathcal{P}^{\Delta_1} + \cdots + \frac{h_{m-1}(0)}{h_m(0)}\mathcal{P}^{\Delta_{m-1}} + \mathcal{P}^{\Delta_m}\right)$$
$$= h_0(0)\mathcal{P}^{\Delta_0} + h_1(0)\mathcal{P}^{\Delta_1} + \cdots + h_{m-1}(0)\mathcal{P}^{\Delta_{m-1}} + h_m(0)\mathcal{P}^{\Delta_m} = 0$$

is a Δ-relation *for all* H_i^*. Since every element $h \in \operatorname{colim}(\mathrm{H}_i^*)$ in the colimit is contained in some $\mathrm{H}_{i_0}^*$ we are done. •

We append the obvious corollary for our construction of the $\mathcal{P}^*$-inseparable closure of an algebra H^*.

COROLLARY 6.1.2: *Let* H^* *be an integral domain of finite* Δ*-length* m*. If each algebra* H_i^* *of the chain of algebras with colimit* $\sqrt[\mathcal{P}^*]{\mathrm{H}^*}$ *has the same* Δ*-length* m*, then* m *is also the* Δ*-length of* $\sqrt[\mathcal{P}^*]{\mathrm{H}^*}$*. Moreover, if we allow coefficients to be taken out of the field of fractions of* H^* *we get a unique normalized* Δ*-relation*

$$\mathbf{t}_0\mathcal{P}^{\Delta_0} + \cdots + \mathbf{t}_m\mathcal{P}^{\Delta_m} = 0,$$

where $\mathbf{t}_m = 1$*, which is valid on* H^**,* $\sqrt[\mathcal{P}^*]{\mathrm{H}^*}$ *and all algebras in between.* •

The following two examples illustrate what happens when we drop various assumptions in the preceding corollary.

EXAMPLE 2: Let $\mathbb{F}[x_1, x_2, x_3, \ldots]$ be the polynomial algebra in infinitely many linear generators, and

$$\mathrm{H}^* := \mathbb{F}[x_1^{p^s}, x_2^{p^s}, x_3^{p^s}, \ldots]$$

be a polynomial ring in infinitely many generators of degree p^s, each being the p^s-th power of a linear form $x_1, x_2, x_3, \ldots$. Then H^* is an integral domain of finite Δ-length $m = 0$, because

$$\mathcal{P}^{\Delta_0} = 0$$

on H^*. H^* has Krull dimension $n = \infty$ and is certainly not Noetherian. We have

$$\mathrm{H}_i^* = \mathbb{F}[x_1^{p^{s-i}}, x_2^{p^{s-i}}, x_3^{p^{s-i}}, \ldots]$$

as you can easily see. Hence as long as $s > i$ the Δ-length of H_i^* remains zero, while for $s = i$ we get

$$\sqrt[\mathcal{P}^*]{\mathrm{H}^*} = \mathrm{H}_s^* = \mathbb{F}[x_1, x_2, x_3, \ldots]$$

which has no finite Δ-length.

EXAMPLE 3 : Let $\mathbb{F}[x_1, x_2, x_3, \ldots]$ be the polynomial algebra in infinitely many linear generators, and

$$\mathrm{H}^* := \mathbb{F}[x_1^p, x_2^{p^2}, x_3^{p^3}, \ldots]$$

be a polynomial ring in infinitely many generators of degree $p, p^2, p^3, \ldots$. Then H^* is a Δ-finite integral domain with Δ-relation

$$\mathcal{P}^{\Delta_0} = 0.$$

We get

$$\mathrm{H}_i^* := \mathbb{F}[x_1, \ldots, x_i, x_{i+1}^p, x_{i+2}^{p^2}, \ldots]$$

and, by Theorem 1.2.3, with Δ-relation

$$(-1)^i \mathbf{d}_{i,0}\mathcal{P}^{\Delta_0} + \cdots + (-1)\mathbf{d}_{i,i-1}\mathcal{P}^{\Delta_{i-1}} + \mathcal{P}^{\Delta_i} = 0.$$

Hence the Δ-length increases by one every time we look at the next algebra in our chain

$$i = m(\mathrm{H}_i^*) < m(\mathrm{H}_{i+1}^*) = i + 1 \quad \forall\, i \geq 0.$$

Moreover the colimit is again

$$\sqrt[\mathcal{P}^*]{\mathrm{H}^*} = \mathbb{F}[x_1, x_2, x_3, \ldots],$$

which has no finite Δ-length.

We turn to Noetherianess.

THEOREM 6.1.3 : *Let* H^* *be an integral domain. Then* H^* *is Noetherian if and only if* $\sqrt[\mathcal{P}^*]{\mathrm{H}^*}$ *is Noetherian.*

PROOF : Since the "if"-part is the contents of Proposition 4.2.3, we deal with the "only if"-part.

Let H^* be Noetherian. We start by collecting some facts we already know:

(1) H_i^* is a Noetherian integral domain, for every $i \geq 0$. This is because the extension $\mathrm{H}^* \hookrightarrow \mathrm{H}_i^*$ is integral and finite[2] and Proposition 4.2.1 part (3).

(2) H_i^* is Δ-finite, $\forall\, i \geq 0$, by Theorem 1.2.1.

(3) The Δ-length of H_i^*, $m_i := m(\mathrm{H}_i^*)$, is less than or equal to the Krull dimension $\dim(\mathrm{H}_i^*) = \dim(\sqrt[\mathcal{P}^*]{\mathrm{H}^*})$, $\forall\, i \geq 0$, by Corollary 1.2.2 and Proposition 4.2.1 (4).

---

[2] Yes, they are finite, because H^* is Noetherian by assumption, recall Section 4.1

CASE 1 : $n := \dim(\mathrm{H}^*) = m(\mathrm{H}^*) =: m$

By Corollary 1.2.2 we have for all $i \in \mathbb{N}_0$ that

$$m := m(\mathrm{H}^*) \leq m(\mathrm{H}_i^*) \leq \dim(\mathrm{H}_i^*) = \dim(\mathrm{H}^*) = n = m.$$

Hence all H_i^* are Δ-finite of the same Δ-length m as H^*. By the preceding Corollary 6.1.2 $\sqrt[\mathcal{P}^*]{\mathrm{H}^*}$ is also Δ-finite of length m with Δ-relation

$$\mathbf{t}_0 \mathcal{P}^{\Delta_0} + \cdots + \mathbf{t}_m \mathcal{P}^{\Delta_m} = 0,$$

where $\mathbf{t}_m = 1$, and we allow coefficients to be taken from the field of fractions of H^*. So the Δ-polynomials for H^* and $\sqrt[\mathcal{P}^*]{\mathrm{H}^*}$ are the same:

$$\Delta(X) = \mathbf{t}_0 X + \cdots + \mathbf{t}_{m-1} X^{q^{m-1}} + \mathbf{t}_m X^{q^m} \in FF(\mathrm{H}^*)[X].$$

We apply the constructions of Chapter 5, in particular those of Section 5.3: let $z_1, \ldots, z_m$ be an $\mathbb{F}_q$-basis for the vector space V consisting of the roots of the common Δ-polynomial of H^* and $\sqrt[\mathcal{P}^*]{\mathrm{H}^*}$. We receive a diagram

$$\begin{array}{ccccc} & & \mathrm{H}^* & \underset{\text{integral}}{\hookrightarrow} & \sqrt[\mathcal{P}^*]{\mathrm{H}^*} \\ & & \downarrow \text{int} & & \downarrow \text{int} \\ \mathbb{F}[z_1, \ldots, z_m] & \hookrightarrow & \mathrm{H}^* < z_1, \ldots, z_m > & \underset{\text{integral}}{\hookrightarrow} & \sqrt[\mathcal{P}^*]{\mathrm{H}^*} < z_1, \ldots, z_m >, \end{array}$$

where the inclusion $\mathrm{H}^* \hookrightarrow \mathrm{H}^* < z_1, \ldots, z_m >$ is integral because H^* is Noetherian by Proposition 5.2.1 part (4), and $\mathrm{H}^* \hookrightarrow \sqrt[\mathcal{P}^*]{\mathrm{H}^*}$ is integral by Proposition 4.2.1 part(4). Since $\sqrt[\mathcal{P}^*]{\mathrm{H}^*}$ is $\mathcal{P}^*$-inseparably closed, so is $\sqrt[\mathcal{P}^*]{\mathrm{H}^*} < z_1, \ldots, z_m >$ by Lemma 5.3.1. We look at the fields of fractions involved

$$\begin{array}{ccccc} \mathbb{F}[z_1, \ldots, z_m] & \hookrightarrow & \mathrm{H}^* < z_1, \ldots, z_m > & \underset{\text{integral}}{\hookrightarrow} & \sqrt[\mathcal{P}^*]{\mathrm{H}^*} < z_1, \ldots, z_m >, \\ \downarrow & & \downarrow & & \downarrow \\ \mathbb{F}(z_1, \ldots, z_m) & \hookrightarrow & FF(\mathrm{H}^* < z_1, \ldots, z_m >) & \hookrightarrow & FF(\sqrt[\mathcal{P}^*]{\mathrm{H}^*} < z_1, \ldots, z_m >). \end{array}$$

Hence, keeping in mind that H^* and $\mathrm{H}^* < z_1, \ldots, z_m >$ are Noetherian, we have

$$\begin{aligned} \operatorname{trzdeg}(\mathbb{F}(z_1, \ldots, z_m)/\mathbb{F}) &= m \\ &= \dim(\mathrm{H}^*) \\ &= \dim(\mathrm{H}^* < z_1, \ldots, z_m >) \\ &= \operatorname{trzdeg}(FF(\mathrm{H}^* < z_1, \ldots, z_m >)/\mathbb{F}), \end{aligned}$$

i.e., the first field extension in the preceding diagram is algebraic. The second field extension is also algebraic, because $\mathrm{H}^* \hookrightarrow \sqrt[\mathcal{P}^*]{\mathrm{H}^*}$ is integral. Therefore Lemma 3.1.1 yields an integral ring extension

$$\mathcal{U}n\left(FF(z_1, \ldots, z_m)\right) = \mathbb{F}[z_1, \ldots, z_m] \hookrightarrow \mathrm{H}^* < z_1, \ldots, z_m > \hookrightarrow \sqrt[\mathcal{P}^*]{\mathrm{H}^*} < z_1, \ldots, z_m > \hookrightarrow \mathcal{U}n\left(FF(\sqrt[\mathcal{P}^*]{\mathrm{H}^*} < z_1, \ldots, z_m >)\right).$$

Applying Theorem 3.2.2 then gives

$$\mathbb{F}[z_1, \ldots, z_m] \hookrightarrow \mathrm{H}^* < z_1, \ldots, z_m > \hookrightarrow \sqrt[\mathcal{P}^*]{\mathrm{H}^*} < z_1, \ldots, z_m >$$

is an isomorphism, and in particular all algebras involved are Noetherian. Therefore

$$\mathrm{H}^* \hookrightarrow \sqrt[\mathcal{P}^*]{\mathrm{H}^*} \hookrightarrow \mathrm{H}^* < z_1, \ldots, z_m >$$

is an integral extension with Noetherian algebras H^* and $\mathrm{H}^* < z_1, \ldots, z_m >$ of the same Krull dimension $n = m$. These the extensions are finite, hence $\sqrt[\mathcal{P}^*]{\mathrm{H}^*}$ is Noetherian.

CASE 2 : $n := \dim(\mathrm{H}^*) > m(\mathrm{H}^*) =: m$

We claim that there exists an index $i_0 \in \mathbb{N}_0$ such that

$$m_{i_0} := m(\mathrm{H}^*_{i_0}) > m(\mathrm{H}^*) =: m.$$

Assuming that we had proven this, we would have, that for some index i_k

$$m(\mathrm{H}^*_{i_k}) = \dim(\mathrm{H}^*_{i_k}) \left(= \dim(\mathrm{H}^*) = n\right).$$

Since $\mathrm{H}^*_{i_k}$ is again Noetherian we could apply Case 1 and we would be done. So what is left to show is that the Δ-length goes up after finitely many steps in our chain of successive ring extensions that leads to $\sqrt[\mathcal{P}^*]{\mathrm{H}^*}$. We assume to the contrary that for every $i \geq 0$

$$m(\mathrm{H}^*_i) = m(\mathrm{H}^*) = m < n = \dim(\mathrm{H}^*).$$

This would give us an infinite chain

$$\mathrm{H}^* = \mathrm{H}^*_0 \subseteq \mathrm{H}^*_1 \subseteq \mathrm{H}^*_2 \subseteq \cdots$$

of Noetherian algebras with the same Δ-length m. By Corollary 6.1.2 the colimit $\sqrt[\mathcal{P}^*]{\mathrm{H}^*}$ has finite Δ-length equals $m < n = \dim(\sqrt[\mathcal{P}^*]{\mathrm{H}^*})$. Recall again the constructions from Chapter 5: If $z_1, \ldots, z_m$ is a basis of the vector space over $\mathbb{F}$ of the roots of the Δ-polynomial of $\sqrt[\mathcal{P}^*]{\mathrm{H}^*}$ then by Lemma 5.3.1

$$\sqrt[\mathcal{P}^*]{\mathrm{H}^*} < z_1, \ldots, z_m >$$

is $\mathcal{P}^*$-inseparably closed. Since

$$\mathrm{H}^* \hookrightarrow \mathrm{H}^* < z_1, \ldots, z_m > \hookrightarrow \sqrt[\mathcal{P}^*]{\mathrm{H}^*} < z_1, \ldots, z_m >$$

are integral ring extensions by Proposition 5.2.1 part (4) and Proposition 4.2.1 part (4), all algebras involved have the same Krull dimension n. However, Theorem 5.3.3 shows that

$$\sqrt[\mathcal{P}^*]{\mathrm{H}^*} < z_1, \ldots, z_m > = \mathbb{F}[z_1, \ldots, z_m],$$

i.e., has dimension $m < n$. This is a contradiction. •

We are now able to prove the promised converse of Theorem 5.3.5.

COROLLARY 6.1.4 : *Let* H^* *be a Noetherian integral domain. Then* H^* *is* $\mathcal{P}^*$*-inseparably closed if and only if its* Δ*-length equals to its Krull dimension* $m(\mathrm{H}^*) = \dim(\mathrm{H}^*)$.

Proof: The "only if" part was proven in Theorem 5.3.5. So, we suppose that $m = m(\mathrm{H}^*) = \dim(\mathrm{H}^*)$. By the preceding Theorem 6.1.3 also $\sqrt[\mathcal{P}^*]{\mathrm{H}^*}$ is Noetherian. Moreover, by Theorem 5.3.5

$$m(\sqrt[\mathcal{P}^*]{\mathrm{H}^*}) = \dim(\sqrt[\mathcal{P}^*]{\mathrm{H}^*}) = m.$$

As above we get a diagram

$$\begin{array}{ccccc} & & \mathrm{H}^* & \xhookrightarrow[\text{int}]{} & \sqrt[\mathcal{P}^*]{\mathrm{H}^*} \\ & & \downarrow \text{int} & & \downarrow \text{int} \\ \mathbb{F}[z_1, \dots, z_m] & \hookrightarrow & \mathrm{H}^* < z_1, \dots, z_m > & \hookrightarrow & \sqrt[\mathcal{P}^*]{\mathrm{H}^*} < z_1, \dots, z_m >, \end{array}$$

where $z_1, \dots, z_m$ are the zeros of the common Δ-polynomial. By Theorem 5.3.5 the algebra $\sqrt[\mathcal{P}^*]{\mathrm{H}^*} < z_1, \dots, z_m >$ is $\mathcal{P}^*$-inseparably closed and hence

$$\mathbb{F}[z_1, \dots, z_m] \hookrightarrow \mathrm{H}^* < z_1, \dots, z_m > \hookrightarrow \sqrt[\mathcal{P}^*]{\mathrm{H}^*} < z_1, \dots, z_m >,$$

is an isomorphism, i.e.,

$$\mathrm{H}^* < z_1, \dots, z_m > \cong \mathbb{F}[z_1, \dots, z_m]$$

is $\mathcal{P}^*$-inseparably closed, and therefore so is H^*, by Theorem 5.3.5. •

We can prove now the complete Embedding Theorem, see Theorem 1.1 of [1] for the case of prime fields.

Corollary 6.1.5 (Embedding Theorem): *Let* H^* *be a Noetherian unstable integral domain of Krull dimension* n. *Then* H^* *can be embedded integrally into a polynomial ring of the same Krull dimension with linear generators.*

Proof: Since H^* is Noetherian, Theorem 6.1.3 tells us that also $\sqrt[\mathcal{P}^*]{\mathrm{H}^*}$ is Noetherian. So, $\sqrt[\mathcal{P}^*]{\mathrm{H}^*}$ is a Noetherian $\mathcal{P}^*$-inseparably closed unstable integral domain and hence can be embedded integrally into a polynomial ring of the desired form by Theorem 5.3.4. Since the extension $\mathrm{H}^* \hookrightarrow \sqrt[\mathcal{P}^*]{\mathrm{H}^*}$ is integral we are done. •

Remark: Again, don't miss the big thing about this theorem: Since H^* and $\mathbb{F}[z_1, \dots, z_n]$ are Noetherian and the inclusion

$$\mathrm{H}^* \hookrightarrow \mathbb{F}[z_1, \dots, z_n]$$

is integral, it is also *finite*, i.e., $\mathbb{F}[z_1, \dots, z_n]$ is finite as a module over H^*.

Remark: If the algebra H^* is not Noetherian, but its $\mathcal{P}^*$-inseparable closure is Δ-finite of length n, then we have a non-integral embedding

$$\mathrm{H}^* \hookrightarrow \sqrt[\mathcal{P}^*]{\mathrm{H}^*} \hookrightarrow \mathbb{F}[z_1, \dots, z_n],$$

by the Embedding Theorem (5.3.4).

6.2 Turkish Delights

After all that hard work we are going to be rewarded with some nice results about the spectrum of $\mathcal{P}^*$-invariant homogeneous prime ideals of H^*. Throughout this section we consider only Noetherian unstable algebras. We start by recalling some terminology.

Consider the spectrum of homogeneous prime ideals in H^*, $\mathcal{Proj}(\mathrm{H}^*)$. We denote by

$$\mathcal{Proj}_{\mathcal{P}^*}(\mathrm{H}^*)$$

the subset of homogeneous prime ideals which are closed under the action of the Steenrod algebra. We call this **the spectrum of $\mathcal{P}^*$-invariant prime ideals of** H^*. See [26], [29], [33], [35] and [36] Chapter 11 for results concerning $\mathcal{P}^*$-invariant ideals.

PROPOSITION **6.2.1** (Turkish Delight 1): *Let H^* be a Noetherian unstable algebra over the Steenrod algebra of Krull dimension n. Then H^* has only finitely many $\mathcal{P}^*$-invariant prime ideals.*

PROOF: Since H^* is Noetherian it has only finitely many prime ideals $\mathfrak{p}_0$ of height 0, which are all $\mathcal{P}^*$-invariant by Theorem 1 in [20] (see also Proposition 11.2.3 in [36] or the more general result in Theorem 3.5 [29]). Therefore $\mathrm{H}^*/\mathfrak{p}_0$ is again an unstable algebra over the Steenrod algebra and moreover an integral domain. Hence by Corollary 6.1.5 there exists an integral extension of unstable algebras over $\mathcal{P}^*$

$$\mathrm{H}^*/\mathfrak{p}_0 \hookrightarrow \mathbb{F}[z_1, \ldots, z_n]$$

where the polynomial ring has linear generators. Since the extension is integral, for any prime ideal $\mathfrak{p} \subset \mathrm{H}^*/\mathfrak{p}_0$, there exists a prime ideal in $\mathbb{F}[z_1, \ldots, z_n]$ lying over it. Since we have an extension of unstable algebras over the Steenrod algebra all prime ideals in $\mathbb{F}[z_1, \ldots, z_n]$ lying over a $\mathcal{P}^*$-invariant prime ideal $\mathfrak{p} \subset \mathrm{H}^*/\mathfrak{p}_0$ are themselves $\mathcal{P}^*$-invariant, see Theorem 2.3 in [26]. By [33] any $\mathcal{P}^*$-invariant prime ideal in $\mathbb{F}[z_1, \ldots, z_n]$ is generated by linear forms. Since there are only finitely many linear forms, there are only finitely many $\mathcal{P}^*$-invariant prime ideals in $\mathbb{F}[z_1, \ldots, z_n]$. Therefore, there are only finitely many $\mathcal{P}^*$-invariant prime ideals in $\mathrm{H}^*/\mathfrak{p}_0$, and hence also in H^*. •

PROPOSITION **6.2.2** (Turkish Delight 2): *Let H^* be a Noetherian unstable algebra over the Steenrod algebra of Krull dimension n. Then, for every $i = 0, \ldots, n$, there exists a $\mathcal{P}^*$-invariant prime ideal in H^* of height i.*

PROOF: By Theorem 1 in [20] (compare Proposition 11.2.3 in [36] or Theorem 3.5 in [29]) all prime ideals $\mathfrak{p}_0 \subset \mathrm{H}^*$ of height 0 are $\mathcal{P}^*$-invariant. Hence our statement is proven for $i = 0$. Proceeding as above, we may divide out by a height zero prime ideal $\mathfrak{p}_0$ and get

$$\mathrm{H}^* \xrightarrow{\mathrm{pr}} \mathrm{H}^*/\mathfrak{p}_0,$$

which is again an unstable algebra over $\mathcal{P}^*$ of the same Krull dimension n. Moreover, it embeds integrally as an algebra over $\mathcal{P}^*$ into a polynomial algebra with linear generators

$$\mathrm{H}^*/\mathfrak{p}_0 \hookrightarrow \mathbb{F}[z_1, \ldots, z_n],$$

by Corollary 6.1.5. In the polynomial algebra, we can find, for any given height, a $\mathcal{P}^*$-invariant prime ideal, e.g., take an ideal from the following saturated maximal chain

$$(0) \subsetneqq (z_1) \subsetneqq (z_1, z_2) \subsetneqq \cdots \subsetneqq (z_1, \ldots, z_n) \subsetneqq \mathbb{F}[z_1, \ldots, z_n].$$

Then

$$(0) \subsetneqq (z_1) \cap \left(\mathrm{H}^*/\mathfrak{p}_0\right) \subsetneqq \cdots \subsetneqq (z_1, \ldots, z_n) \cap \left(\mathrm{H}^*/\mathfrak{p}_0\right) \subsetneqq \mathrm{H}^*/\mathfrak{p}_0$$

is a saturated maximal chain of $\mathcal{P}^*$-invariant prime ideals in $\mathrm{H}^*/\mathfrak{p}_0$ by Lemma 2.1 in [26]: This is because our extension of algebras is integral, so in particular satisfies lying over, compare Section 3.1 in [4]. This implies that

$$\mathfrak{p}_0 \subsetneqq \mathrm{pr}^{-1}\left((z_1) \cap \mathrm{H}^*/\mathfrak{p}_0\right) \subsetneqq \cdots \subsetneqq \mathrm{pr}^{-1}\left((z_1, \ldots, z_n) \cap \mathrm{H}^*/\mathfrak{p}_0\right) \subsetneqq \mathrm{H}^*$$

is a saturated maximal chain of $\mathcal{P}^*$-invariant prime ideals in H^*. •

The following proposition justifies the remark in the introduction that the spectrum of $\mathcal{P}^*$-invariants homogeneous prime ideals , $\mathcal{P}roj_{\mathcal{P}^*}(\mathrm{H}^*)$, forms a chain saturated poset (with respect to inclusion).

PROPOSITION 6.2.3 (Turkish Delights 3 and 4): *Let H^* be a Noetherian unstable algebra over the Steenrod algebra of Krull dimension n. Let $\mathfrak{p} \subset \mathrm{H}^*$ be a $\mathcal{P}^*$-invariant prime ideal of height i, $i = 0, \ldots, n$. Then*

(1) *there exists an ascending saturated maximal chain of $\mathcal{P}^*$-invariant prime ideals in H^* starting with $\mathfrak{p}$ and ending with the maximal ideal $\mathfrak{m}$.*

(2) *there exists a descending saturated maximal chain of $\mathcal{P}^*$-invariant prime ideals in H^* starting at $\mathfrak{p}$ and ending with a $\mathcal{P}^*$-invariant prime ideal $\mathfrak{p}_0$ of height 0.*

PROOF: Let $\mathfrak{p}$ have height i. Note that if $i = 0$ the statement of the proposition is proven in Turkish Delight 2. Otherwise, take a prime ideal of height 0 sitting inside our given $\mathfrak{p}$,

$$\mathfrak{p}_0 \subsetneqq \mathfrak{p} \subset \mathrm{H}^*,$$

and divide it out, to obtain

$$\mathrm{pr} : \mathrm{H}^* \longrightarrow \mathrm{H}^*/\mathfrak{p}_0.$$

Our choice of $\mathfrak{p}_0$ guarantees that $\mathfrak{p}/\mathfrak{p}_0 \subset \mathrm{H}^*/\mathfrak{p}_0$ is still a $\mathcal{P}^*$-invariant prime ideal of height i. Again, we embed the quotient algebra in a polynomial ring with linear generators

$$\mathrm{H}^*/\mathfrak{p}_0 \hookrightarrow \mathbb{F}[z_1, \ldots, z_n].$$

Since this embedding is integral, the lying over and going up theorems hold. So, take a prime ideal

$$\mathfrak{q} \subset \mathbb{F}[z_1, \ldots, z_n]$$

lying over our given $\mathfrak{p}/\mathfrak{p}_0$, i.e.,

$$\mathfrak{q} \cap (\mathrm{H}^*/\mathfrak{p}_0) = \mathfrak{p}/\mathfrak{p}_0.$$

Then $\mathfrak{q}$ is also $\mathcal{P}^*$-invariant by Theorem 2.3 in [26]. By [33] we can find saturated ascending and descending maximal chains of $\mathcal{P}^*$-invariant prime ideals in the polynomial ring staring with $\mathfrak{q}$ and ending with the maximal, resp. staring with $\mathfrak{q}$ and ending at (0):

$$\mathfrak{q} \subsetneq \cdots \subsetneq \mathfrak{m} \subset \mathbb{F}[z_1, \ldots, z_n]$$

and

$$(0) \subsetneq \cdots \subsetneq \mathfrak{q} \subset \mathbb{F}[z_1, \ldots, z_n].$$

Intersect these chains with $\mathrm{H}^*/\mathfrak{p}_0$ and pull them back to H^* via pr^{-1}, and we are done. •

6.3 Noetherianess II

In this section we generalize Theorem 6.1.3 to reduced algebras H^*.

This can be done in two different ways: On the one hand we could exploit the fact that Theorem 4.3.1 gives a bijection between the homogeneous prime spectra $\mathcal{P}roj(\mathrm{H}^*)$ and $\mathcal{P}roj(\sqrt[\mathcal{P}^*]{\mathrm{H}^*})$. On the other hand we could use Proposition 4.2.4, where we have proven that $\sqrt[\mathcal{P}^*]{\mathrm{H}^*}$ is Noetherian if and only if this colimit is a finite union.

Anyway, let's state the theorem first:

THEOREM 6.3.1: *Let* H^* *be a reduced, unstable algebra. Then* H^* *is Noetherian if and only if* $\sqrt[\mathcal{P}^*]{\mathrm{H}^*}$ *is Noetherian.*

PROOF: The "if"-part is the contents of Proposition 4.2.3. So we have to prove the "only if"-part, and we assume that H^* is Noetherian.

Since H^* is Noetherian, there are only finitely many, say k, homogeneous prime ideals $\mathfrak{p}_1, \ldots, \mathfrak{p}_k \subset \mathrm{H}^*$ of height zero. By a result of Peter S. Landweber, Theorem 1 in [20] or Proposition 11.2.3 in [36] or, more generally, Theorem 3.5 in [29], they are all $\mathcal{P}^*$-invariant. Hence $\mathrm{H}^*/\mathfrak{p}_i$ is again an unstable Noetherian algebra, for all $i = 1, \ldots, k$, and moreover an integral domain. So, its $\mathcal{P}^*$-inseparable closure $\sqrt[\mathcal{P}^*]{\mathrm{H}^*/\mathfrak{p}_i}$ is Noetherian for all $i = 1, \ldots, k$ by Theorem 6.1.3. Denote by $\mathfrak{q}_i \subset \sqrt[\mathcal{P}^*]{\mathrm{H}^*}$ the *unique* (by Theorem 4.3.1) homogeneous prime ideal in $\sqrt[\mathcal{P}^*]{\mathrm{H}^*}$ lying over $\mathfrak{p}_i$. By Proposition 4.3.5 we have

$$\mathrm{H}^*/\mathfrak{p}_i \subseteq \sqrt[\mathcal{P}^*]{\mathrm{H}^*}/\mathfrak{q}_i \subseteq \sqrt[\mathcal{P}^*]{\mathrm{H}^*/\mathfrak{p}_i}, \quad \forall\, i = 1, \ldots, k.$$

We know that the smallest and the biggest ring are Noetherian and that the extension is integral. Therefore $\sqrt[\mathcal{P}^*]{\mathrm{H}^*/\mathfrak{p}_i}$ is finitely generated as a module over $\mathrm{H}^*/\mathfrak{p}_i$. Hence so is $\sqrt[\mathcal{P}^*]{\mathrm{H}^*}/\mathfrak{q}_i$. So $\sqrt[\mathcal{P}^*]{\mathrm{H}^*}/\mathfrak{q}_i$ is Noetherian for all i, and hence so is

$$\sqrt[\mathcal{P}^*]{\mathrm{H}^*} \Big/ \left(\bigcap_{i=1}^{k} \mathfrak{q}_i \right).$$

The ideals $\mathfrak{q}_1, \ldots, \mathfrak{q}_k \subset \sqrt[\mathcal{P}^*]{\mathrm{H}^*}$ are all of height zero, and, again by Theorem 4.3.1, there are no other homogeneous prime ideals in $\sqrt[\mathcal{P}^*]{\mathrm{H}^*}$ of height zero, therefore

$$\bigcap_{i=1}^{k} \mathfrak{q}_i = \mathcal{N}il\left(\sqrt[\mathcal{P}^*]{\mathrm{H}^*}\right) = (0),$$

by Proposition 4.2.1 part (2), and hence

$$\sqrt[\mathcal{P}^*]{\mathrm{H}^*} \Big/ \left(\bigcap_{i=1}^{k} \mathfrak{q}_i \right) = \sqrt[\mathcal{P}^*]{\mathrm{H}^*}$$

is Noetherian. •

Alternative Proof: For a prime ideal $\mathfrak{p} \subset \mathrm{H}^*$ denote by

$$\mathrm{pr}_{\mathfrak{p}} : \mathrm{H}^* \longrightarrow \mathrm{H}^*/\mathfrak{p}$$

the canonical projection onto the quotient. Since $\mathcal{N}il(\mathrm{H}^*) = (0)$ by assumption we have an embedding

$$L := \bigoplus_{i=1}^{k} \mathrm{pr}_i : \mathrm{H}^* \hookrightarrow \bigoplus_{i=1}^{k} \mathrm{H}^*/\mathfrak{p}_i$$

where the direct sum runs over all height zero prime ideals $\mathfrak{p}_i$ of H^* and the i-th component of L is projection onto $\mathrm{H}^*/\mathfrak{p}_i$. This is a finite direct sum, because H^* is Noetherian.

Applying the construction of the $\mathcal{P}^*$-inseparable closure gives a pair of shifts

$$\begin{array}{ccc}
\mathrm{H}^* & \overset{L}{\hookrightarrow} & \bigoplus_{i=1}^{k} \mathrm{H}^*/\mathfrak{p}_i \\
\downarrow & & \downarrow \\
\mathrm{H}^*_1 & & \left(\bigoplus_{i=1}^{k} \mathrm{H}^*/\mathfrak{p}_i \right)_1 \\
& & \| \\
& & \bigoplus_{i=1}^{k} (\mathrm{H}^*/\mathfrak{p}_i)_1 \\
\downarrow & & \downarrow \\
\vdots & & \vdots
\end{array}$$

By Proposition 4.2.1 the algebra H_1 is again reduced and Noetherian and therefore also admits an embedding

$$L_1 := \overset{k_1}{\underset{i=1}{\oplus}} \mathrm{pr}_i : \mathrm{H}_1^* \hookrightarrow \bigoplus_{i=1}^{k_1} \mathrm{H}_1^*/\mathfrak{q}_i.$$

By Theorem 4.3.1 $k_1 = k$ and by Lemma 4.3.4

$$L_1 := \oplus_{i=1}^{k} \mathrm{pr}_i : \mathrm{H}_1^* \hookrightarrow \bigoplus_{i=1}^{k} \mathrm{H}_1^*/\mathfrak{q}_i \hookrightarrow \bigoplus_{i=1}^{k} \left(\mathrm{H}^*/\mathfrak{p}_i\right)_1.$$

Hence we can complete our diagram above as follows:

$$\begin{array}{ccc} \mathrm{H}^* & \overset{L}{\hookrightarrow} & \bigoplus_{i=1}^{k} \mathrm{H}^*/\mathfrak{p}_i \\ \downarrow & & \downarrow \\ \mathrm{H}_1^* & \overset{L_1}{\hookrightarrow} & \bigoplus_{i=1}^{k} \left(\mathrm{H}^*/\mathfrak{p}_i\right)_1. \end{array}$$

By Theorem 6.1.3 the $\mathcal{P}^*$-inseparable closure $\sqrt[\mathcal{P}^*]{\mathrm{H}^*/\mathfrak{p}_i}$ is Noetherian, for every i. Therefore, by Proposition 4.2.4 the colimits involved are finite unions

$$\sqrt[\mathcal{P}^*]{\mathrm{H}^*/\mathfrak{p}_i} := \operatorname{colim}_j \left((\mathrm{H}^*/\mathfrak{p}_i)_j\right) = \bigcup_{j=0}^{r_i} \left((\mathrm{H}^*/\mathfrak{p}_i)_j\right) = \left(\mathrm{H}^*/\mathfrak{p}_i\right)_{r_i}$$

for some $r_i \in \mathbb{N}_0$. Set $r := \max\{r_1, \dots, r_k\}$ then we have

$$\begin{array}{ccc} \mathrm{H}^* & \overset{L}{\hookrightarrow} & \bigoplus_{i=1}^{k} \mathrm{H}^*/\mathfrak{p}_i \\ \downarrow & & \downarrow \\ \mathrm{H}_1^* & \overset{L_1}{\hookrightarrow} & \bigoplus_{i=1}^{k} \left(\mathrm{H}^*/\mathfrak{p}_i\right)_1 \\ \downarrow & & \downarrow \\ \mathrm{H}_2^* & \overset{L_2}{\hookrightarrow} & \bigoplus_{i=1}^{k} \left(\mathrm{H}^*/\mathfrak{p}_i\right)_2 \\ \downarrow & & \downarrow \\ \vdots & & \vdots \\ \downarrow & & \downarrow \\ \mathrm{H}_r^* & \overset{L_r}{\hookrightarrow} & \bigoplus_{i=1}^{k} \left(\mathrm{H}^*/\mathfrak{p}_i\right)_r \\ \downarrow & & \| \\ \mathrm{H}_{r+1}^* & \overset{L_{r+1}}{\hookrightarrow} & \bigoplus_{i=1}^{k} \left(\mathrm{H}^*/\mathfrak{p}_i\right)_{r+1} \\ & & \| \\ & & \sqrt[\mathcal{P}^*]{\bigoplus_{i=1}^{k} \mathrm{H}^*/\mathfrak{p}_i} \end{array}$$

We claim that $H_r^* = H_{r+1}^*$, and hence by Theorem 4.2.4 $\sqrt[\mathcal{P}^*]{H^*}$ is Noetherian. To this end take an element $h \in H_{r+1}^*$. Denote by

$$\mathfrak{q}_{r,1}, \ldots, \mathfrak{q}_{r,k} \subset H_r^*$$

the minimal prime ideals of H_r^* and by

$$\mathfrak{q}_{r+1,1}, \ldots, \mathfrak{q}_{r+1,k} \subset H_{r+1}^*$$

those of H_{r+1}^*. Without loss of generality we can assume that

$$\mathfrak{q}_{r,j} = \mathfrak{q}_{r+1,j} \cap H_r^* \quad \forall j = 1, \ldots, k.$$

For all $j = 1, \ldots, k$, consider the commutative diagrams

$$\begin{array}{ccc} H_r^* & \overset{\mathrm{pr}_{r,j}}{\longrightarrow} & H_r^*/\mathfrak{q}_{r,j} \\ \downarrow & & \| \\ H_{r+1}^* & \overset{\mathrm{pr}_{r+1,j}}{\longrightarrow} & H_{r+1}^*/\mathfrak{q}_{r+1,j}. \end{array}$$

Then for every $h \in H_{r+1}^*$ we have

$$\mathrm{pr}_{r,j}^{-1}\left(\mathrm{pr}_{r+1,j}(h)\right) - h \in \mathfrak{q}_{r+1,j}$$

for any $j = 1, \ldots, k$. Therefore

$$\mathrm{pr}_{r,j}^{-1}\left(\mathrm{pr}_{r+1,j}(h)\right) - h \in \mathfrak{q}_{r+1,1} \cap \cdots \cap \mathfrak{q}_{r+1,k} = \mathcal{N}il(H_{r+1}^*) = (0)$$

and hence

$$\mathrm{pr}_{r,j}^{-1}\left(\mathrm{pr}_{r+1,j}(h)\right) = h$$

as was to be shown. •

Chapter 7
The Galois Embedding Theorem, the Little Imbedding Theorem, and A Bit More

After all the work we have done so far we are prepared to take the remaining steps towards the Galois Embedding Theorem. We will assume in the first section of this chapter that our H^* is an $\mathcal{P}^*$-inseparably closed, integrally closed unstable integral domain. After that, we will have a look at what we can do when H^* is not $\mathcal{P}^*$-inseparably closed. Then we will have a look at algebras which contain integrally the Dickson algebra (or a fractal thereof). We will determine *the* Δ-relation for H^*, find that the copy of the (fractal of the) Dickson algebra in H^* is often unique and the H^*-module $\Delta(\mathrm{H}^*)$ free. This will help us to prove the Little Imbedding Theorem, which says that we can always find a fractal of the Dickson algebra in an unstable integral domain H^*.

Throughout the whole chapter H^* is an unstable algebra over the Steenrod algebra.

7.1 The Galois Embedding Theorem

Assume H^* to be an integrally closed, $\mathcal{P}^*$-inseparably closed, Noetherian unstable integral domain over $\mathbb{F}$.

Firstly, we are in the lucky position that we can embed H^* integrally into a polynomial algebra with linear generators by Theorem 6.1.5:

$$\mathrm{H}^* \hookrightarrow \mathbb{F}[V].$$

Hence we have an algebraic extension of the fields of fractions

$$FF(\mathrm{H}^*) \hookrightarrow \mathbb{F}(V) = \mathbb{E}^*.$$

This extension is separable, because $\mathbb{E}^*$ is the splitting field of the separable Δ-polynomial $\Delta(X)$. Therefore this field extension is also finite and normal, see [48] Theorem 13 on page 76. So, our extension is Galois with some Galois group $G \leq \mathrm{Aut}(\mathbb{F}(V))$, i.e.,

$$FF(\mathrm{H}^*) = \mathbb{F}(V)^G,$$

and we have integral extensions

$$\mathrm{H}^* \hookrightarrow \mathbb{F}[V]^G \hookrightarrow \mathbb{F}[V]$$

of integrally closed domains, where the first two have the same field of fractions. Hence they must be equal and we have proved the generalization of Theorem 1.2 in [1] to arbitrary Galois fields.

THEOREM 7.1.1 (Galois Embedding Theorem): *Let H^* be an unstable algebra over $\mathbb{F}$. Then H^* is a ring of invariants $\mathbb{F}[V]^G$ if and only if H^* is an integrally closed, $\mathcal{P}^*$-inseparably closed, Noetherian integral domain.* •

7.2 A Bit More

In this section H^* remains a Noetherian unstable integral domain over $\mathbb{F}$.

An essential assumption in the preceding section was that H^* is $\mathcal{P}^*$-inseparably closed. If this is not the case then at the level of fields of fractions the extension

$$FF(\mathrm{H}^*) \hookrightarrow \mathbb{F}(V) = \mathbb{E}^*$$

is not separable. As in classical field theory we have two possibilities[1] to bring order into this mess: the first is to find an intermediate field $\mathbb{K}^*$

$$FF(\mathrm{H}^*) \underset{\mathcal{P}^*\text{–purely insep.}}{\hookrightarrow} \mathbb{K}^* \underset{\text{sep.}}{\hookrightarrow} \mathbb{F}(V)$$

such that the first extension is $\mathcal{P}^*$-purely inseparable and the second separable. We then get:

THEOREM 7.2.1: *Let H^* be a Noetherian unstable integral domain. The field $\mathbb{K}^*$ is the field of fractions of the $\mathcal{P}^*$-inseparable closure $\sqrt[\mathcal{P}^*]{\mathrm{H}^*}$ of H^*. Moreover, if H^* is integrally closed, then $\sqrt[\mathcal{P}^*]{\mathrm{H}^*}$ is the integral closure of H^* in $\mathbb{K}^*$. In this case $\sqrt[\mathcal{P}^*]{\mathrm{H}^*}$ fulfills the assumption of the Galois Embedding Theorem, so is a ring of invariants.*

PROOF: From Corollary 2.3.4 we get a $\mathcal{P}^*$-purely inseparable field extension

$$\mathbb{K}^* \subseteq FF(\mathrm{H}^*)_{\mathcal{P}^*\text{–insep}} \subsetneqq \mathbb{F}(V)_{\mathcal{P}^*\text{–insep}} = \mathbb{F}(V).$$

Therefore

$$\mathbb{K}^* = FF(\mathrm{H}^*)_{\mathcal{P}^*\text{–insep}}.$$

Then the first statement follows from Proposition 4.2.6, namely $\mathbb{K}^* = FF(\sqrt[\mathcal{P}^*]{\mathrm{H}^*})$. Therefore we have the following situation:

$$\begin{array}{ccccc} \mathrm{H}^* & \hookrightarrow & \sqrt[\mathcal{P}^*]{\mathrm{H}^*} & \hookrightarrow & \mathbb{F}[V] \\ \downarrow & & \downarrow & & \downarrow \\ FF(\mathrm{H}^*) & \underset{\mathcal{P}^*\text{–purely insep.}}{\hookrightarrow} & FF(\sqrt[\mathcal{P}^*]{\mathrm{H}^*}) = \mathbb{K}^* & \underset{\text{sep.}}{\hookrightarrow} & \mathbb{F}(V) \end{array}$$

[1] Recall the remarks at the end of Section 2.3.

where the first field extension is $\mathcal{P}^*$-purely inseparable, while the last one is separable by Proposition 4.2.6. By Proposition 4.2.1 part (5), $\sqrt[\mathcal{P}^*]{\mathrm{H}^*}$ is integrally closed whenever H^* is. Hence

$$\overline{\mathrm{H}^*}_{FF(\sqrt[\mathcal{P}^*]{H})} \subseteq \sqrt[\mathcal{P}^*]{\mathrm{H}^*}$$

is an integral extension of integrally closed domains with the same field of fractions, so they are equal.

Finally $\sqrt[\mathcal{P}^*]{\mathrm{H}^*}$ is an integrally closed, $\mathcal{P}^*$-inseparably closed Noetherian unstable integral domain. Hence $\sqrt[\mathcal{P}^*]{\mathrm{H}^*}$ is a ring of invariants. That's what we wanted. •

Let's have an example to illustrate this.

EXAMPLE 1: Consider the the polynomial $\mathbb{F}$-algebra $\mathbb{F}[x, y]$ generated by two linear forms, and take the subalgebra generated[2] by x^{2p}, $x^p y$ and y^2. We have

$$\begin{array}{ccc} \mathbb{F} < x^{2p},\ x^p y,\ y^2 > & \hookrightarrow & \mathbb{F}[x,\ y] \\ \downarrow & & \downarrow \\ \mathbb{F}(x^{2p},\ x^p y) & \hookrightarrow & \mathbb{F}(x,\ y). \end{array}$$

The $\mathcal{P}^*$-inseparable closure $\mathbb{K}^*$ of the field of fractions $\mathbb{F}(x^{2p},\ x^p y)$ of the smaller ring inside $\mathbb{F}(x,\ y)$ is obviously

$$\mathbb{K}^* = \mathbb{F}(x^2,\ xy),$$

and the integral closure of the little *algebra* in $\mathbb{F}(x,\ y)$ is

$$\overline{\mathbb{F} < x^{2p},\ x^p y,\ y^2 >}_{\mathbb{F}(x,\ y)} = \mathbb{F}[x^2,\ y^2,\ xy] \Big/ \left((xy)^2 - x^2 y^2\right).$$

This is, as our theorem predicts, indeed a ring of invariants, namely of the 2-dimensional $\mathbb{Z}/2$ representation afforded by the matrix

$$\begin{bmatrix} -1 & 0 \\ 0 & -1 \end{bmatrix}.$$

On the other hand, we can find another intermediate field $\mathbb{L}^*$ such that

$$FF(\mathrm{H}^*) \underset{\text{sep.}}{\hookrightarrow} \mathbb{L}^* \underset{\mathcal{P}^*\text{-purely insep.}}{\hookrightarrow} \mathbb{F}(V).$$

This time the first extension is separable, and the second $\mathcal{P}^*$-purely inseparable. Recall the notation

$$\mathbb{F}[V] := \mathbb{F}[z_1, \ldots, z_n].$$

We have (compare also [46]):

[2] Right, you've guessed it: $p \neq 2$.

THEOREM 7.2.2 : *Let H^* be a Noetherian unstable integral domain. The integral closure of H^* in $\mathbb{L}^*$, denoted by $\overline{H^*_{\mathbb{L}^*}}$, is the polynomial algebra*

$$\mathbb{F}[z_1^{p^{e_1}}, \ldots, z_n^{p^{e_n}}]$$

for some p-th powers $p^{e_1}, \ldots, p^{e_n}$. Moreover, if H^* is integrally closed

$$H^* = \left(\overline{H^*_{\mathbb{L}^*}}\right)^G = \mathbb{F}[z_1^{p^{e_1}}, \ldots, z_n^{p^{e_n}}]^G$$

for a subgroup $G \leq GL(n, \mathbb{F})$, where the action of the general linear group on $\overline{H^*_{\mathbb{L}^*}}$ is induced by the natural one on $\mathbb{F}[V]$.

PROOF: First note that $\overline{H^*_{\mathbb{L}^*}}$ is an integrally closed domain with field of fractions $\mathbb{L}^*$. Therefore we have

$$\begin{array}{ccccc} H^* & \hookrightarrow & \overline{H^*_{\mathbb{L}^*}} & \hookrightarrow & \mathbb{F}[V] = \mathbb{F}[z_1, \ldots, z_n] \\ \downarrow & & \downarrow & & \downarrow \\ FF(H^*) & \underset{\text{sep.}}{\hookrightarrow} & FF\left(\overline{H^*_{\mathbb{L}^*}}\right) = \mathbb{L}^* & \underset{\mathcal{P}^*\text{-purely insep.}}{\hookrightarrow} & \mathbb{F}(V) = \mathbb{F}(z_1, \ldots, z_n) \end{array}$$

where at the algebra level we have integral extensions, and, at the field level, the first one is separable, and the second one is $\mathcal{P}^*$-purely inseparable. Hence for any z_i, $i = 1, \ldots, n$ there exists a p-th power p^{e_i} such that

$$z_i^{p^{e_i}} \in \mathbb{L}^*.$$

Since, in addition, $z_i^{p^{e_i}}$ is integral over $\overline{H^*_{\mathbb{L}^*}}$, we have

$$\begin{array}{ccccc} \mathbb{F}[z_1^{p^{e_1}}, \ldots, z_n^{p^{e_n}}] & \hookrightarrow & \overline{H^*_{\mathbb{L}^*}} & \hookrightarrow & \mathbb{F}[V] \\ \downarrow & & \downarrow & & \downarrow \\ \mathbb{F}(z_1^{p^{e_1}}, \ldots, z_n^{p^{e_n}}) & \hookrightarrow & FF\left(\overline{H^*_{\mathbb{L}^*}}\right) = \mathbb{L}^* & \hookrightarrow & \mathbb{F}(V), \end{array}$$

where at the algebra level we have again integral extensions of integrally closed domains, and, at field level, $\mathcal{P}^*$-purely inseparable extensions. If we choose the exponents p^{e_i} to be minimal with this property we get a separable extension

$$FF(H^*) \subseteq \mathbb{F}(z_1^{p^{e_1}}, \ldots, z_n^{p^{e_n}}).$$

Since we started with a separable extension

$$FF(H^*) \subseteq \mathbb{L}^*$$

the field extension

$$\mathbb{F}(z_1^{p^{e_1}}, \ldots, z_n^{p^{e_n}}) \subseteq \mathbb{L}^*$$

is also separable. Hence these two fields must be equal, and, therefore

$$\mathbb{F}[z_1^{p^{e_1}}, \ldots, z_n^{p^{e_n}}] = \overline{H^*_{\mathbb{L}^*}}.$$

Hence H^* is $\mathcal{P}^*$-inseparably closed in $\overline{H^*_{\mathbb{L}^*}}$. The field extension

$$FF(H^*) \subset \mathbb{F}(z_1^{p^{e_1}}, \ldots, z_n^{p^{e_n}})$$

is also normal, because its the splitting field of $\Delta(X)$. If we assume in addition that H^* is integrally closed the rest of the statement follows from the Galois Embedding Theorem 7.1.1 •

Let's have a second look at Example 1.

EXAMPLE 2: Recall

$$\begin{array}{ccc} \mathbb{F} < x^{2p},\ x^p y,\ y^2 > & \hookrightarrow & \mathbb{F}[x,\ y] \\ \downarrow & & \downarrow \\ \mathbb{F}(x^{2p},\ x^p y) & \hookrightarrow & \mathbb{F}(x,\ y). \end{array}$$

This time we are looking for the separable closure of the little field, resp. algebra, in the big field. We find

$$\mathbb{L}^* = \mathbb{F}(x^p,\ y) \text{ and } \overline{\mathbb{F} < x^{2p},\ x^p y,\ y^2 >_{\mathbb{F}(x,\ y)}} = \mathbb{F}[x^p,\ y]$$

and so

$$\mathbb{F} < x^{2p},\ x^p y,\ y^2 > = \mathbb{F}[x^p,\ y]^{\mathbb{Z}/2}.$$

7.3 Uniqueness Theorems

In this section we show that, for a fixed q-th power, an integrally embedded fractal of the Dickson algebra inside H^* is unique provided H^* is reduced, Δ-finite and has only finitely many height zero homogeneous prime ideals. So, H^* is reduced, Δ-finite and has only finitely many height zero homogeneous prime ideals throughout the whole section.

Recall from the introduction that a fractal of $\mathcal{D}^*(n)$ is

$$\mathcal{D}^*(n)^{q^s} = \mathbb{F}[\mathbf{d}_{n,0}^{q^s}, \ldots, \mathbf{d}_{n,n-1}^{q^s}] = \mathbb{F}[z_1^{q^s}, \ldots, z_n^{q^s}]^{\mathrm{GL}(n,\ \mathbb{F})}.$$

PROPOSITION 7.3.1: *Let* H^* *be a reduced* Δ*-finite unstable algebra. Assume that there are only finitely many height zero homogeneous prime ideals in* H^**. If* H^* *contains integrally the Dickson algebra,* $\mathcal{D}^*(n)$*, where* $n = \dim(\mathrm{H}^*)$*, then the* Δ*-relation for* H^* *is*

$$(-1)^n \mathbf{d}_{n,0} \mathcal{P}^{\Delta_0} + \cdots + (-1)\mathbf{d}_{n,n-1}\mathcal{P}^{\Delta_{n-1}} + \mathcal{P}^{\Delta_n} = 0.$$

Moreover, H^* *is Noetherian.*

PROOF:

CASE 1 : If H^* is an integral domain, then we have maps

$$\begin{array}{ccc} \mathcal{D}^*(n) & \underset{\text{int}}{\hookrightarrow} & \mathrm{H}^* \\ \downarrow \text{integral} & & \downarrow \\ \mathbb{F}[z_1, \ldots, z_n] & \hookrightarrow & \mathrm{H}^* < z_1, \ldots, z_n >. \end{array}$$

Since $\mathbb{F}[z_1, \ldots, z_n]$ is integral over the Dickson algebra $\mathcal{D}^*(n)$, it is integral over H^*. Therefore also $\mathbb{F}[z_1, \ldots, z_n] \hookrightarrow \mathrm{H}^* < z_1, \ldots, z_n >$ is integral, and therefore by Theorem 3.2.2 an isomorphism. Hence we have a chain of integral extensions

$$\mathcal{D}^*(n) \hookrightarrow \mathrm{H}^* \hookrightarrow \mathbb{F}[z_1, \ldots, z_n] =: \mathbb{F}[V].$$

Since the Δ-relation on $\mathbb{F}[V]$ is of the desired form, the one for H^* is also (compare the remarks after Theorem 1.2.3). Moreover, since $\mathbb{F}[V]$ is finite over $\mathcal{D}^*(n)$, so is H^* and therefore H^* is Noetherian.

CASE 2 : If H^* is reduced, then take a height zero homogeneous prime ideal $\mathfrak{p} \subset \mathrm{H}^*$ and form the quotient $\mathrm{H}^*/\mathfrak{p}$. This is again an unstable Δ-finite algebra, and moreover, since $\mathcal{D}^*(n) \hookrightarrow \mathrm{H}^*$ is integral, we have

$$\begin{array}{ccc} \mathcal{D}^*(n) & \underset{\text{int}}{\hookrightarrow} & \mathrm{H}^* \\ \downarrow \text{int} & \swarrow & \\ \mathrm{H}^*/\mathfrak{p} & & \end{array}$$

By Case 1 the Δ-relation on $\mathrm{H}^*/\mathfrak{p}$ is

$$\left((-1)^n \mathbf{d}_{n,0}\mathcal{P}^{\Delta_0} + \cdots + (-1)\mathbf{d}_{n,n-1}\mathcal{P}^{\Delta_{n-1}} + \mathcal{P}^{\Delta_n}\right) = 0.$$

In other words, for any $\mathfrak{p} \in \mathcal{P}roj(\mathrm{H}^*)$ of height zero

$$(-1)^n \mathbf{d}_{n,0}\mathcal{P}^{\Delta_0} + \cdots + (-1)\mathbf{d}_{n,n-1}\mathcal{P}^{\Delta_{n-1}} + \mathcal{P}^{\Delta_n} \in \mathfrak{p}.$$

Therefore

$$(-1)^n \mathbf{d}_{n,0}\mathcal{P}^{\Delta_0} + \cdots + (-1)\mathbf{d}_{n,n-1}\mathcal{P}^{\Delta_{n-1}} + \mathcal{P}^{\Delta_n} \in \bigcap_{\mathfrak{p} \in \mathcal{P}roj(\mathrm{H}^*),\ \mathcal{h}t(\mathfrak{p})=0} \mathfrak{p} = \mathcal{N}il(\mathrm{H}^*) = (0),$$

which was to be shown. Finally, since $\mathrm{H}^*/\mathfrak{p}$ is Noetherian by case 1 and there are only finitely many height zero prime ideals, by assumption, H^* is also Noetherian by Exercise 1 in Section 2.1 of [4]. •

We can extend this to the case where we have only a fractal of the Dickson algebra in H^*.

COROLLARY 7.3.2 : *Let H^* be a reduced Δ-finite unstable algebra with finitely many height zero homogeneous prime ideals. If H^* contains integrally a fractal, $\mathcal{D}^*(n)^{q^s}$, of the Dickson algebra and $n = \dim(\mathrm{H}^*)$, then*

$$(-1)^n \mathbf{d}_{n,0}^{q^s}\mathcal{P}^{\Delta_s} + \cdots + (-1)\mathbf{d}_{n,n-1}^{q^s}\mathcal{P}^{\Delta_{n-1+s}} + \mathcal{P}^{\Delta_{n+s}} = 0$$

on H^*. Moreover, H^* is Noetherian.

PROOF: If $\mathcal{D}^*(n)^{q^s} \hookrightarrow \mathrm{H}^*$ is an integral extension, then

$$\mathcal{D}^*(n) \hookrightarrow \sqrt[\mathcal{P}^*]{\mathrm{H}^*}$$

is integral. Therefore Proposition 7.3.1 tells us that the Δ-relation for $\sqrt[\mathcal{P}^*]{\mathrm{H}^*}$ is

$$(-1)^n \mathbf{d}_{n,0}\mathcal{P}^{\Delta_0} + \cdots + (-1)\mathbf{d}_{n,n-1}\mathcal{P}^{\Delta_{n-1}} + \mathcal{P}^{\Delta_n} = 0.$$

By Lemma 1.1.8 we have that

$$(-1)^n \mathbf{d}_{n,0}^{q^s} \mathcal{P}^{\Delta_s} + \cdots + (-1)\mathbf{d}_{n,n-1}^{q^s} \mathcal{P}^{\Delta_{n-1+s}} + \mathcal{P}^{\Delta_{n+s}} = 0$$

on $\sqrt[\mathcal{P}^*]{\mathrm{H}^*}$ and a fortiori on H^*. Since $\sqrt[\mathcal{P}^*]{\mathrm{H}^*}$ is Noetherian by Proposition 7.3.1 so is H^* by Proposition 4.2.3. •

REMARK: The preceding results tell us that a reduced unstable algebra H^* is Noetherian if it is Δ-finite, has only finitely many height zero prime ideals and contains integrally a Dickson algebra. We will see in Proposition 8.2.1 that these conditions are also necessary.

REMARK: Let H^* be reduced, Δ-finite with finitely many height zero prime ideals. Note that we now know that there is an integral extension $\mathcal{D}^*(n) \hookrightarrow \mathrm{H}^*$ if and only if on H^* the following Δ-relation holds

$$(-1)^n \mathbf{d}_{n,0} \mathcal{P}^{\Delta_0} + \cdots + (-1)\mathbf{d}_{n,n-1} \mathcal{P}^{\Delta_{n-1}} + \mathcal{P}^{\Delta_n} = 0.$$

The "if"-part follows from Theorem 5.1.8. The "only if"-part is given by the above proposition.

The uniqueness of an integrally embedded Dickson algebra (or a fractal thereof) is now easy.

THEOREM 7.3.3 (Uniqueness Theorem): *Let H^* be a reduced Δ-finite unstable algebra. with finitely many height zero prime ideals. For $i = 1, 2$ denote by $\mathcal{D}^*(n)_i^{q^s}$ a fractal of the Dickson algebra of Krull dimension n and let*

$$\varphi_i : \mathcal{D}^*(n)_i^{q^s} \hookrightarrow \mathrm{H}^*$$

be integral extensions. Then

$$\mathcal{D}^*(n)_1^{q^s} = \mathcal{D}^*(n)_2^{q^s}.$$

PROOF: We consider first the case where $s = 0$, i.e., we have Dickson algebras inside H^*. Denote by $\mathbf{d}_{n,j}(i)$ the Dickson classes of $\mathcal{D}^*(n)_i$ for $i = 1, 2$ and $j = 0, \ldots, n-1$. Proposition 7.3.1 leads to two Δ-relations on H^*, namely

$$(-1)^n \mathbf{d}_{n,0}(i) \mathcal{P}^{\Delta_0} + \cdots + (-1)\mathbf{d}_{n,n-1}(i) \mathcal{P}^{\Delta_{n-1}} + \mathcal{P}^{\Delta_n} = 0$$

for $i = 1, 2$. By minimality of $n = m(\mathrm{H}^*)$ the coefficients must coincide.

Next we consider the general case, where $s > 0$. We embed H^* into its $\mathcal{P}^*$-inseparable closure $\sqrt[\mathcal{P}^*]{\mathrm{H}^*}$. We compose this embedding with the maps φ_i, for $i = 1, 2$

$$\mathcal{D}^*(n)_i^{q^s} \overset{\varphi_i}{\hookrightarrow} \mathrm{H}^* \hookrightarrow \sqrt[\mathcal{P}^*]{\mathrm{H}^*}, \quad i = 1, 2.$$

Since $\sqrt[\mathcal{P}^*]{\mathrm{H}^*}$ is $\mathcal{P}^*$-inseparably closed, the $\mathcal{P}^*$-inseparable closure of $\mathcal{D}^*(n)_1^{q^s}$, resp. of $\mathcal{D}^*(n)_2^{q^s}$, i.e., the Dickson algebra itself, is contained in $\sqrt[\mathcal{P}^*]{\mathrm{H}^*}$, so we get integral extensions

$$\Phi_i : \mathcal{D}^*(n)_i \hookrightarrow \sqrt[\mathcal{P}^*]{\mathrm{H}^*}, \quad i = 1,\ 2.$$

By the first part of this proof, these two copies of the Dickson algebra must coincide

$$\mathcal{D}^*(n)_1 = \mathcal{D}^*(n)_2 \subset \sqrt[\mathcal{P}^*]{\mathrm{H}^*}.$$

Hence the fractals are also equal

$$\mathcal{D}^*(n)_1^{q^s} = \mathcal{D}^*(n)_2^{q^s} \subset \mathrm{H}^*.$$

That's what we wanted. •

Remark: If we drop the condition on H^* that $\mathcal{N}il(\mathrm{H}^*) = (0)$, but assume that the nil radical is nilpotent, we get that for any pair of fractals of the Dickson algebra $\mathcal{D}^*(n)_i^{q^{s_i}}$, $i = 1,\ 2$, large enough q-th powers coincide

$$\mathcal{D}^*(n)_1^{q^l} = \mathcal{D}^*(n)_2^{q^l}$$

for an $l \geq s_1,\ s_2$. To be precise: suppose without loss of generality $s = s_1 \geq s_2$. Then the preceding theorem tells us that

$$\mathcal{D}^*(n)_1^{q^s} = \mathcal{D}^*(n)_2^{q^s} \subset \mathrm{H}^*/\mathcal{N}il(\mathrm{H}^*).$$

Let $l \in \mathbb{N}_0$, $l \geq s$, such that

$$\mathcal{N}il(\mathrm{H}^*)^{q^l} = (0).$$

Then pulling back the q^l-th fractals of the two Dickson algebras to H^* leads to

$$\mathcal{D}^*(n)_1^{q^l} = \mathcal{D}^*(n)_2^{q^l} + \mathcal{N}il(\mathrm{H}^*)^{q^l} = \mathcal{D}^*(n)_2^{q^l} \subset \mathrm{H}^*$$

as claimed.

7.4 The Little Imbedding Theorem

An obvious consequence of the Galois Embedding Theorem is that the Dickson algebra $\mathcal{D}^*(n) \hookrightarrow \mathrm{H}^*$ embeds integrally into an H^* which fulfills all the requirements of that theorem. However, this is almost true if we just assume that H^* is an unstable Noetherian integral domain: H^* always contains integrally a fractal of the Dickson algebra.

We first show that we can reduce the situation to $\mathcal{P}^*$-inseparably closed algebras.

Lemma 7.4.1: *Let* $\sqrt[\mathcal{P}^*]{\mathrm{H}^*}$ *contain integrally a fractal of the Dickson algebra* $\mathcal{D}^*(n)$*, say* $\mathcal{D}^*(n)^{q^s}$ *for some* $s \geq 0$*. Then* H^* *contains a fractal of the Dickson algebra, namely* $\mathcal{D}^*(n)^{q^l}$ *for some* $l \geq s \geq 0$.

PROOF: By assumption we have the following diagram

$$\begin{array}{ccc} \mathrm{H}^* & \hookrightarrow & \sqrt[\mathcal{P}^*]{\mathrm{H}^*} \\ & & \uparrow \\ & & \mathcal{D}^*(n)^{q^s} \end{array}$$

Since a certain p-th power of any element in $\sqrt[\mathcal{P}^*]{\mathrm{H}^*}$ is inside H^* we can find natural numbers $l_0, \ldots, l_{n-1}$ such that

$$\left(\mathbf{d}_{n,i}^{q^s}\right)^{q^{l_i}} \in \mathrm{H}^* \ \forall\, i = 0, \ldots, n-1.$$

If we choose $l \in \mathbb{N}_0$ to be the maximum of $s + l_0, \ldots, s + l_{n-1}$ then

$$\mathcal{D}^*(n)^{q^l} \subseteq \mathrm{H}^*.$$

•

Here is another technical lemma, that makes life easier later on:

LEMMA 7.4.2: *Let H^* be a $\mathcal{P}^*$-inseparably closed Noetherian unstable integral domain of Krull dimension n. Consider the embedding*

$$\mathrm{H}^* \hookrightarrow \mathbb{F}[V]$$

given by the results of Section 5.3. If H^* contains some power of the top Dickson class

$$\mathbf{d}_{n,0}^{\alpha} \in \mathrm{H}^* \quad \alpha \in \mathbb{N}_0$$

then it contains $\mathbf{d}_{n,0}$ itself.

PROOF: By assumption we have that

$$\mathbf{d}_{n,0}^{\alpha} = \left(\prod_{v \in V^* \setminus 0} v\right)^{\alpha} \in \mathrm{H}^*,$$

for some $\alpha \in \mathbb{N}_0$. Without loss of generality assume α to be minimal with this property, and assume that $\alpha > 1$. Then p does not divide α, because otherwise

$$\mathbf{d}_{n,0}^{\frac{\alpha}{p}} \in \mathrm{H}^*$$

since H^* is $\mathcal{P}^*$-inseparably closed. We express α as a multiple of p with some remainder, i.e., let

$$\alpha = kp + r,$$

for some $k,\ r \in \mathbb{N}_0$, $0 < r < p$. We get

$$\mathcal{P}^{\Delta_n}(\mathbf{d}_{n,0}^{\alpha}) = \alpha \mathbf{d}_{n,0}^{\alpha+1},$$

where we used Proposition A.2.1. So, applying $\mathcal{P}^{\Delta_n}$ $(p-r)$-times to $\mathbf{d}_{n,0}^{\alpha}$ gives

$$\begin{aligned} \mathcal{P}^{\Delta_n} \cdots \mathcal{P}^{\Delta_n}(\mathbf{d}_{n,0}^{\alpha}) &= \alpha(\alpha+1)\cdots(\alpha+r-1)\mathbf{d}_{n,0}^{\alpha+p-r} \\ &= \alpha(\alpha+1)\cdots(\alpha+r-1)\mathbf{d}_{n,0}^{(k+1)p}. \end{aligned}$$

The coefficient is non-zero modulo p by construction. So, it follows that

$$\mathbf{d}_{n,0}^{k+1} \in \mathrm{H}^*,$$

because H^* is $\mathcal{P}^*$-inseparably closed. As $k+1 < \alpha$, unless $\alpha = 1$, this contradicts the minimality of α. Hence

$$\mathbf{d}_{n,0} \in \mathrm{H}^*,$$

as claimed. •

The set up of the Galois Embedding Theorem hands us a copy of the Dickson algebra in any $\mathcal{P}^*$-inseparably closed, integrally closed, integral domain. The next result shows that it is enough to assume that H^* is a $\mathcal{P}^*$-inseparably closed integral domain, to conclude that H^* contains a copy of the top Dickson class, which implies the Little Imbedding Theorem for the case of algebras of Krull dimension 1 (compare Theorem 1.1 on page 24 of [19] for prime fields).

PROPOSITION 7.4.3: *Let H^* be a Noetherian unstable integral domain of Krull dimension n. Consider the embedding*

$$\mathrm{H}^* \hookrightarrow \mathbb{F}[V]$$

given by the Embedding Theorem 6.1.5. Then H^* contains a q-th power of the top Dickson class $\mathbf{d}_{n,0} \in \mathbb{F}[V]$. If in addition H^* is $\mathcal{P}^*$-inseparably closed then

$$\mathbf{d}_{n,0} \in \mathrm{H}^*.$$

PROOF: By Lemma 7.4.1 we need only consider $\mathcal{P}^*$-inseparably closed integral domains H^*. We proceed by induction on the Krull dimension n.

If $n = 0$, then $\mathrm{H}^* = \mathbb{F} = \mathcal{D}^*(0)$ and nothing needs to be shown.

If $n = 1$, then the Galois Embedding Theorem 7.1.1 hands us a copy of the Dickson algebra in the integral closure of H^*, $\overline{\mathrm{H}^*}$

$$\begin{array}{ccccc} & & \mathrm{H}^* & & \\ & & \downarrow & & \\ \mathbb{F}[x^{q-1}] & \hookrightarrow & \overline{\mathrm{H}^*} & \hookrightarrow & \mathbb{F}[x]. \end{array}$$

Let $x^\alpha \in \mathrm{H}^*$ be a non-zero element of minimal degree $\alpha \in \mathbb{N}$. Then

$$\mathbf{d}_{1,0}^{\alpha} = \left(x^{q-1}\right)^{\alpha} = \left(x^{\alpha}\right)^{q-1} \in \mathrm{H}^*.$$

Hence $\mathbf{d}_{1,0} = x^{q-1} \in \mathrm{H}^*$ by Lemma 7.4.2.

For the general case, assume that $n > 1$ and the result is established of the Krull dimension of H^* is $\leq n-1$. Take the embedding of H^* into a polynomial ring $\mathbb{F}[V]$ given by Embedding Theorem, Corollary 6.1.5. Choose a linear form $l \in \mathbb{F}[V]$, and consider the projections

$$\begin{array}{ccc} \mathrm{H}^* & \hookrightarrow & \mathbb{F}[V] \\ \downarrow & & \downarrow \\ \mathrm{H}^*/((l) \cap \mathrm{H}^*) & \hookrightarrow & \mathbb{F}[W] \end{array}$$

where $\operatorname{span}_{\mathbb{F}}(W,\ l) = V$. The ideal $(l) \subset \mathbb{F}[V]$ is $\mathcal{P}^*$-invariant prime of height one. Hence so is the contraction $(l) \cap \mathrm{H}^* \subset \mathrm{H}^*$, by Lemma 2.1 in [26]. Therefore $\mathbb{F}[W]$ and $\mathrm{H}^*/((l) \cap \mathrm{H}^*)$ are unstable integral domains of Krull dimension $n-1$. We apply the induction hypothesis and get that a q-th power of the top Dickson class in $\mathbb{F}[W]$ is also in $\mathrm{H}^*/((l) \cap \mathrm{H}^*)$

$$\mathbf{d}_{n-1,0}^{q^{\beta(l)}}(l) \in \mathrm{H}^*/((l) \cap \mathrm{H}^*).$$

The preimage of $\mathbf{d}_{n-1,0}^{q^{\beta(l)}}(l)$ under the projection is

$$\mathrm{pr}^{-1}\left(\mathbf{d}_{n-1,0}^{q^{\beta(l)}}(l)\right) = \left(\mathbf{d}_{n-1,0}^{q^{\beta(l)}}(l)\right) + \left((l) \cap \mathrm{H}^*\right) \ni \left(\mathbf{d}_{n-1,0}^{q^{\beta(l)}}(l)\right) + 0 = \left(\prod_{v \in W^* \setminus \{0\}} v\right)^{q^{\beta(l)}}.$$

We play the same game for all linear forms $l \in \mathbb{F}[V]$, set $\beta = \max\{\beta(l)\}$ and multiply all the inverse images together we get:

$$\begin{aligned}
\prod_{l \in V^* \setminus 0} \mathbf{d}_{n-1,0}^{q^{\beta}}(l) &= \prod_{l \in V^* \setminus 0} \left(\prod_{v \in W^* \setminus \{0\},\ \operatorname{Span}_{\mathbb{F}}(W,\ l) = V} v\right)^{q^{\beta}} \\
&= \prod_{W^* < V^*,\ \dim_{\mathbb{F}}(W^*) = n-1} \left(\prod_{v \in W^* \setminus 0} v\right)^{(q^n - q^{n-1})q^{\beta}} \\
&= \left(\prod_{v \in V^* \setminus 0} v\right)^{\alpha} \quad \text{for some } \alpha \in \mathbb{N}_0,
\end{aligned}$$

where the last equation follows by symmetry. Hence we have that

$$\mathbf{d}_{n,0}^{\alpha} = \left(\prod_{v \in V^* \setminus 0} v\right)^{\alpha} \in \mathrm{H}^*,$$

for some $\alpha \in \mathbb{N}_0$. Applying Lemma 7.4.2 gives

$$\mathbf{d}_{n,0} \in \mathrm{H}^*,$$

as claimed. •

THEOREM 7.4.4 (Little Imbedding Theorem): *Let* H^* *be a Noetherian unstable integral domain. Then there exists a natural number* $l \in \mathbb{N}_0$, *and an inclusion*

$$\mathcal{D}^*(n)^{q^l} \hookrightarrow \mathrm{H}^*$$

which is an integral extension of unstable algebras over $\mathcal{P}^*$. *Moreover, if* H^* *is in addition* $\mathcal{P}^*$*-inseparably closed, then* $l = 0$.

PROOF: By Lemma 7.4.1 it suffices to consider the case where H^* is $\mathcal{P}^*$-inseparably closed.

Consider first a $\mathcal{P}^*$-inseparably closed unstable integral domain H^* which is also integrally closed. Then we have, by the Galois Embedding Theorem 7.1.1, that H^* is a ring of invariants, i.e.,

$$\mathcal{D}^*(n) \hookrightarrow \mathrm{H}^* = \mathbb{F}[V]^G \hookrightarrow \mathbb{F}[V]$$

for some finite group G and nothing needs to be shown.

Consider the case of a $\mathcal{P}^*$-inseparably closed integral domain H^* which is not integrally closed. Denote by $\overline{\mathrm{H}^*}$, the integral closure of H^*. From the above, we can assume that there is a Dickson algebra $\mathcal{D}^*(n)$ inside $\overline{\mathrm{H}^*}$. We have

$$\begin{array}{ccc} \mathcal{D}^*(n) & \underset{\text{finite}}{\hookrightarrow} & \overline{\mathrm{H}^*} \subseteq \mathbb{F}[V] \\ \uparrow & & \uparrow \\ \mathcal{D}^*(n) \cap \mathrm{H}^* & \hookrightarrow & \mathrm{H}^* \end{array}$$

where every extension is integral.

We want to proceed by induction on the Krull dimension n. If $n = 0$ then our diagram looks like

$$\begin{array}{ccc} \mathbb{F} & \hookrightarrow & \mathbb{F} \subseteq \mathbb{F} \\ \uparrow & & \uparrow \\ \mathbb{F} \cap \mathrm{H}^* & \hookrightarrow & \mathbb{F} \end{array}$$

and this implies

$$\mathcal{D}^*(0) \cap \mathrm{H}^* = \mathbb{F} \cap \mathrm{H}^* = \mathbb{F} = \mathcal{D}^*(0)$$

because H^* is connected. For $n = 1$ the result is the contents of Proposition 7.4.3. So, let's suppose that $n > 1$. Let $l \in \mathbb{F}[V]$ be a linear form. Then by taking the quotient mod l in the above diagram leads, by Proposition A.3.2, to

$$\begin{array}{ccc} & \mathcal{D}^*(n-1)^q & \underset{\text{finite}}{\hookrightarrow} \overline{\mathrm{H}^*}/\left((l) \cap \overline{\mathrm{H}^*}\right) \\ & & \nearrow \text{integral} \\ \mathrm{H}^*/\left((l) \cap \mathrm{H}^*\right) & & \end{array}$$

The quotient $\mathrm{H}^*/\left((l) \cap \mathrm{H}^*\right)$ is a Noetherian integral domain, but of Krull dimension one less. Therefore the induction hypothesis hands us a fractal of the Dickson algebra in the quotient

$$\check{\mathcal{D}}^*(n-1)^{q^\beta} \underset{\text{finite}}{\hookrightarrow} \mathrm{H}^*/\left((l) \cap \mathrm{H}^*\right) \hookrightarrow \overline{\mathrm{H}^*}/\left((l) \cap \overline{\mathrm{H}^*}\right).$$

By the Uniqueness Theorem 7.3.3 we have

$$\mathcal{D}^*(n-1)^{q^\beta} = \check{\mathcal{D}}^*(n-1)^{q^\beta}.$$

Therefore we can complete our diagram as follows

$$\begin{array}{ccc} \mathcal{D}^*(n-1)^{q^\beta} & \underset{\text{finite}}{\hookrightarrow} & \overline{\mathrm{H}^*}/\left((l)\cap\overline{\mathrm{H}^*}\right) \\ \downarrow \text{finite} & \nearrow \text{finite} & \\ \mathrm{H}^*/\left((l)\cap\mathrm{H}^*\right). & & \end{array}$$

Choose $x \in \mathcal{D}^*(n)$ but not in H^*, i.e., not in $\mathcal{D}^*(n) \cap \mathrm{H}^*$, of minimal positive degree. We want to show that x is in the kernel of the projection map. Assume to the contrary that x is not in the kernel. Then

$$\begin{array}{ccc} x \in \mathcal{D}^*(n-1)^q & \hookrightarrow & \overline{\mathrm{H}^*}/\left((l)\cap\overline{\mathrm{H}^*}\right) \\ \uparrow & & \uparrow \\ \mathcal{D}^*(n-1)^{q^\beta} & \hookrightarrow & \mathrm{H}^*/\left((l)\cap\mathrm{H}^*\right), \end{array}$$

so that

$$x^{q^{\beta-1}} \in \mathcal{D}^*(n-1)^{q^\beta} \subseteq \mathrm{H}^*/\left((l)\cap\mathrm{H}^*\right).$$

This implies that

$$x^{q^{\beta-1}} \in \mathrm{H}^* \hookrightarrow \overline{\mathrm{H}^*}.$$

Since

$$\mathcal{P}^{\Delta_i}(x^{q^{\beta-1}}) = 0 \in \mathcal{D}^*(n) \subseteq \overline{\mathrm{H}^*}$$

this remains true in the subring H^*. Since H^* is $\mathcal{P}^*$-inseparably closed this means that $x \in \mathrm{H}^*$. However, this contradicts the assumption that the element x is not in the intersection of H^* and $\mathcal{D}^*(n)$. Hence we may assume that the element x lies in the kernel of the projection map. Therefore

$$x \in (\mathbf{d}_{n,0}) \in \mathcal{D}^*(n),$$

so

$$x = x'\mathbf{d}_{n,0} \in \mathcal{D}^*(n).$$

Since x is of minimal positive degree we get that $x' \in \mathrm{H}^*$. Hence

$$x = x'\mathbf{d}_{n,0} \in (\mathbf{d}_{n,0}) \cap \mathcal{D}^*(n) \cap \mathrm{H}^*,$$

because Proposition 7.4.3 tells us that the top Dickson class is in H^*. That's a contradiction. •

COROLLARY 7.4.5: *An unstable Noetherian integral domain* H^* *contains a Thom class* $\mathbf{t}$.

PROOF: Obvious, since a large enough power of the top Dickson class is contained in the fractal of the Dickson algebra which we just found inside H^*, $\mathbf{d}_{n,0}^{q^l} \in \mathrm{H}^*$, and Proposition A.2.1 in the Appendix tells us that the top Dickson class (and any of its q-th powers) is a Thom class . •

COROLLARY 7.4.6: *An unstable Noetherian integral domain* H^* *of Krull dimension* n *contains a homogeneous system of parameters,* $h_0, \ldots, h_{n-1}$*, such that the polynomial subalgebra it generates,* $\mathbb{F}[h_0, \ldots, h_{n-1}]$ *is an unstable algebra over the Steenrod algebra.*

PROOF: The (fractal of the) Dickson algebra we just found inside H^* hands us the desired system of parameters: the Dickson classes $\mathbf{d}_{n,0}, \ldots, \mathbf{d}_{n,n-1}$ (or the appropriate q-th powers). •

We finish this chapter with an example of an integral domain H^* which is not $\mathcal{P}^*$-inseparably closed, but still contains a copy of the Dickson algebra, i.e., the converse of the second statement of the Little Imbedding Theorem, Theorem 7.4.4, is not true.

EXAMPLE 1: Consider the field $\mathbb{F} = \mathbb{F}_2$ with two elements, and take a polynomial algebra in two linear generators x, y, $\mathbb{F}[x, y]$. The Dickson algebra in this case is

$$\mathcal{D}^*(2) = \mathbb{F}[x^2y + xy^2,\ x^2 + y^2 + xy] \hookrightarrow \mathbb{F}[x, y].$$

Define the algebra H^* by

$$H^* := \mathbb{F}[x^2 + y^2,\ xy,\ xy(x+y)] / \left((x^2 + y^2)(xy) + (xy(x+y))^2\right).$$

Then H^* is an integral domain containing integrally $\mathcal{D}^*(2)$, but it is not $\mathcal{P}^*$-inseparably closed, because

$$\mathcal{P}^{\Delta_i}(x^2 + y^2) = 0 \quad \forall\ i \geq 0,$$

and

$$x + y \notin H^*.$$

Note, that we have at the level of fields of fractions separable field extensions of $\mathcal{P}^*$-inseparably closed fields:

$$\begin{array}{ccccc} FF(\mathcal{D}^*(2)) & \hookrightarrow & FF(H^*) & \hookrightarrow & FF(\mathbb{F}[x, y]) \\ \| & & \| & & \| \\ \mathbb{F}(x^2y + xy^2,\ xy(x+y)) & \hookrightarrow & \mathbb{F}(x+y,\ xy) & \hookrightarrow & \mathbb{F}(x, y). \end{array}$$

Hence even though the algebra H^* is not $\mathcal{P}^*$-inseparably closed, its field of fractions is, and in this particular case a copy of the Dickson algebra $\mathcal{D}^*(2)$ even sits integrally inside H^*. Finally note that the integral closure of H^* is

$$\mathbb{F}[x + y,\ xy],$$

i.e., a $\mathcal{P}^*$-inseparably closed integrally closed Noetherian unstable integral domain, and, as the Galois Embedding Theorem, 7.1.1, predicts a ring of invariants, namely of the regular representation of $\mathbb{Z}/2$ afforded by the matrix

$$\begin{bmatrix} 0 & 1 \\ 1 & 0 \end{bmatrix} \in GL(2,\ \mathbb{F}).$$

Chapter 8
The Big Imbedding Theorem, Thom Classes, Turkish Delights II and the Reverse Landweber-Stong Conjecture

In this chapter we will drop the assumption that our algebra H^* is an integral domain, and we will show that one of the most important results of the preceding chapters remains true, namely, we can still find a fractal of the Dickson algebra integrally inside H^*. From this we can conclude that any H^* contains a Thom class, which in turn leads to further gorgeous results about the $\mathcal{P}^*$-invariant prime spectrum of H^* (Turkish Delights II), and, to a tool for proofs by induction in this category. Finally, we will give a counter example to the Reverse Landweber-Stong Conjecture.

Throughout the whole chapter H^* remains an unstable algebra over the Steenrod algebra.

8.1 The Big Imbedding Theorem

In this section we prove that we can find a fractal of the Dickson algebra (integrally, of course,) in any unstable Noetherian algebra H^*, even if we drop the condition that H^* be an integral domain.

So, in this section H^* is a Noetherian unstable algebra.

Recall that the Dickson algebra has the following fractal property (see Propositions A.3.1 and A.3.2 in the appendix):

$$\begin{array}{ccc} \mathcal{D}^*(n)^{q^l} & \hookrightarrow & \mathbb{F}[z_1, \ldots, z_n] \\ \downarrow & & \downarrow \text{pr} \\ \mathcal{D}^*(n-1)^{q^{l+1}} & \hookrightarrow & \mathbb{F}[z_2, \ldots, z_n] \end{array}$$

where

$$\mathcal{D}^*(n-1)^{q^{l+1}} = \mathcal{D}^*(n)^{q^l} \Big/ \left((z_1) \cap \mathcal{D}^*(n)^{q^l}\right) = \mathcal{D}^*(n)^{q^l} \Big/ (\mathbf{d}_{n,0}^{q^l}).$$

First we are going to show that, we can restrict our attention to reduced algebras H^*.

The following lemma[1] is due to Larry Smith, [38].

LEMMA 8.1.1: *Let* H^* *be Noetherian. If* $\mathrm{H}^*/\mathcal{N}il(\mathrm{H}^*)$ *contains integrally a fractal of the Dickson algebra then so does* H^*.

PROOF: Consider the Frobenius map

$$\Phi : \mathrm{H}^* \longrightarrow \mathrm{H}^*,\ h \longmapsto h^q.$$

(Obviously this map does not preserve degrees, but this is no bother to us at the moment.) The image $\Phi(\mathrm{H}^*) = \mathrm{H}^{*q} \subseteq \mathrm{H}^*$ is a subalgebra. Moreover the Cartan formulae imply

$$\mathcal{P}^k(\Phi(h)) = \begin{cases} \Phi\left(\mathcal{P}^{\frac{k}{q}}(h)\right) & \text{if } q \mid k \\ 0 & \text{otherwise.} \end{cases}$$

So, in fact $\Phi(\mathrm{H}^*) \subseteq \mathrm{H}^*$ is an unstable subalgebra. Since H^* is Noetherian the nil radical $\mathcal{N}il(\mathrm{H}^*)$ is a nilpotent ideal, i.e., there is an integer $s \in \mathbf{N}_0$ such that

$$\left(\mathcal{N}il(\mathrm{H}^*)\right)^s = (0).$$

Choose $s = q^t$, so that Φ^t annihilates $\mathcal{N}il(\mathrm{H}^*)$ and therefore $\Phi^t(\mathrm{H}^*)$ has no nilpotent elements.

Returning to our problem, take a fractal of the Dickson algebra inside the reduced $\mathrm{H}^*/\mathcal{N}il(\mathrm{H}^*)$

$$\mathcal{D}^*(n)^{q^l} \subseteq \mathrm{H}^*/\mathcal{N}il(\mathrm{H}^*)$$

pull it back to H^* via the canonical projection map,

$$\mathcal{D}^*(n)^{q^l} + \mathcal{N}il(\mathrm{H}^*) = \mathrm{pr}^{-1}\left(\mathcal{D}^*(n)^{q^l}\right) \subseteq \mathrm{H}^*$$

and push it forward[2] to H^{*q^t} via the iterated Frobenius Φ^t,

$$\Phi^t\left(\mathcal{D}^*(n)^{q^l} + \mathcal{N}il(\mathrm{H}^*)\right) \subseteq \mathrm{H}^{*q^t}.$$

Since Φ^t is an additive map we get

$$\mathcal{D}^*(n)^{q^{l+t}} = \Phi^t\left(\mathcal{D}^*(n)^{q^l} + \mathcal{N}il(\mathrm{H}^*)\right) \subseteq \mathrm{H}^{*q^t} \subseteq \mathrm{H}^*$$

and that's what we wanted. •

This lemma allows us to restrict our attention to unstable algebras H^* with trivial nil radical.

---

[1] The following lemma remains true if we replace Noetherian by the condition that the nil radical is nilpotent.

[2] Surely, the Dickson algebra doesn't enjoy being pulled in the one direction and pushed in the other. However, that's how life is.

By Lemma 7.4.1 we can also assume, without loss of generality, that in addition, H^* is $\mathcal{P}^*$-inseparably closed.

So, assume all this; H^* is reduced $\mathcal{P}^*$-inseparably closed. Let $\{\mathfrak{p}_1, \ldots, \mathfrak{p}_k\}$ be the set of all height zero prime ideals of H^*, then

$$(0) = \mathfrak{p}_1 \cap \cdots \cap \mathfrak{p}_k.$$

Recall that we have an embedding

$$L := \oplus_{i=1}^k \mathrm{pr}_i : \mathrm{H}^* \hookrightarrow \bigoplus_{i=1}^k \mathrm{H}^*/\mathfrak{p}_i.$$

The k factors on the right hand side are unstable integral domains of the same Krull dimension n as H^*. Therefore the Little Imbedding Theorem, Theorem 7.4.4, tells us that there are fractals of the Dickson algebra in these factors

$$\mathcal{D}^*(n)_i^{q^{\lambda(i)}} \subseteq \mathrm{H}^*/\mathfrak{p}_i, \ \forall\, i = 1, \ldots, k.$$

Hence we have, for $\lambda := \max\{\lambda(i)\}$,

$$\begin{array}{cccc} L: & \mathrm{H}^* & \hookrightarrow & \bigoplus_{i=1}^k \mathrm{H}^*/\mathfrak{p}_i \\ & & & \uparrow \\ & & & \bigoplus_{i=1}^k \mathcal{D}^*(n)_i^{q^\lambda}, \end{array}$$

and what we are looking[3] for is a common lift of all these fractals to a fractal in H^*. Since we assumed that H^* is $\mathcal{P}^*$-inseparably closed this would give us a Dickson algebra proper inside H^*. Denote, in analogy to the notation used above, $\mathbf{d}_{n,0}^{q^\lambda}(i) \in \mathcal{D}^*(n)_i^{q^\lambda}$ the appropriate power of the top Dickson class in the Dickson fractal $\mathcal{D}^*(n)_i^{q^\lambda} \subset \mathrm{H}^*/\mathfrak{p}_i$.

LEMMA 8.1.2: *With the above notation we have*

$$\mathbf{d}_{n,0}^{q^s}(j) \in \bigcap_{i=1,\ i \neq j}^{k} \mathfrak{p}_i$$

for a suitably large $s \in \mathbb{N}_0$.

PROOF: Since we have integral extensions

$$\mathcal{D}^*(n)_j^{q^\lambda} \hookrightarrow \mathrm{H}^*/\mathfrak{p}_j$$

---

[3] It should be possible to employ Rector's Proposition 3.1, [30], resp. Lam's Corollary 3.4, [19], to achieve this goal. An outline of how this might work is in the Appendix of [6].

any $\mathcal{P}^*$-invariant homogeneous prime ideal $\mathfrak{p} \subset \mathrm{H}^*/\mathfrak{p}_j$ contracts to a $\mathcal{P}^*$-invariant homogeneous prime ideal, $\mathfrak{p} \cap \mathcal{D}^*(n)_j^{q^\lambda}$, in the Dickson algebra (by Lemma 2.1 in [26]). Therefore, for a suitable $l \in \{1, \ldots, n-1\}$,

$$\mathfrak{p} \cap \mathcal{D}^*(n)_j^{q^\lambda} = \left(\mathbf{d}_{n,0}^{q^\lambda}(j), \ldots, \mathbf{d}_{n,l}^{q^\lambda}(j)\right) \subset \mathcal{D}^*(n)_j$$

by (a straightforward generalization of) Theorem 1 in [20] (see also [36] Theorem 11.4.6). In particular, the fractal of our top Dickson class $\mathbf{d}_{n,0}^{q^\lambda}(j)$ is an element of any non-trivial $\mathcal{P}^*$-invariant homogeneous prime ideal

$$\mathbf{d}_{n,0}^{q^\lambda}(j) \in \mathfrak{p} \cap \mathcal{D}^*(n)_j^{q^\lambda} \subseteq \mathfrak{p}.$$

This means that for any non-trivial $\mathcal{P}^*$-invariant radical ideal $I = \mathcal{R}ad(I) \subset \mathrm{H}^*/\mathfrak{p}_j$

$$\mathbf{d}_{n,0}^{q^\lambda}(j) \in \mathcal{R}ad(I) \cap \mathcal{D}^*(n)_j^{q^\lambda}.$$

In turn, this implies, that for any non-trivial $\mathcal{P}^*$-invariant ideal $I \subset \mathrm{H}^*/\mathfrak{p}_j$, there is an integer $s(I) \in \mathbb{N}_0$ such that

$$\left(\mathbf{d}_{n,0}^{q^\lambda}(j)\right)^{q^{s(I)}} \in I \cap \mathcal{D}^*(n)_j^{q^\lambda} \subset \mathcal{D}^*(n)_j^{q^\lambda} \hookrightarrow \mathrm{H}^*/\mathfrak{p}_j.$$

In particular, for $s_j = \lambda + \max\{s(\mathfrak{p}_0), \ldots, \widehat{s(\mathfrak{p}_j)}, \ldots, s(\mathfrak{p}_k)\}$,

$$\left(\mathbf{d}_{n,0}(j)\right)^{q^{s_j}} \in \mathfrak{p}_i/\mathfrak{p}_j \subset \mathrm{H}^*/\mathfrak{p}_j$$

for any $i = 1, \ldots, k,\ i \neq j$, because when $\mathfrak{p}_i$ is $\mathcal{P}^*$-invariant so is $\mathfrak{p}_i/\mathfrak{p}_j$. And therefore

$$\left(\mathbf{d}_{n,0}(j)\right)^{q^{s_j}} \in \bigcap_{i=1,\ i\neq j}^{k} \mathfrak{p}_i/\mathfrak{p}_j \subset \mathrm{H}^*/\mathfrak{p}_j,$$

i.e.,

$$\left(\mathbf{d}_{n,0}(j)\right)^{q^{s_j}} \in \bigcap_{i=1,\ i\neq j}^{k} \mathfrak{p}_i \subset \mathrm{H}^*.$$

Set $s = \max\{s_1, \ldots, s_k\}$ and we are done. •

REMARK: Note that $\mathbf{d}_{n,0}^{q^{s_j}}(j)$ generates a height zero $\mathcal{P}^*$-invariant principal ideal in H^*, because it is an element in all but one height zero prime ideal of H^*. In particular it follows that $\mathbf{d}_{n,0}^{q^{s_j}}(j)$ is a "universal" zero divisor in H^* for any $j = 0, \ldots, k$. [4]

Let's have an example of this.

[4] Of course, all this looks a bit odd if H^* is an integral domain. So to make sense out of this, you should keep in mind that we are looking at algebras with zero divisors, so there are really non-trivial prime ideals of height zero.

Example 1: Let

$$\mathrm{H}^* = \mathbb{F}[x,\ y]/(xy)$$

be the $\mathbb{F}$-algebra generated by two linear forms x and y with the single relation $xy = 0$. Our embedding looks like

$$\begin{array}{cccc} L: & \mathbb{F}[x,\ y]/(xy) & \hookrightarrow & \mathbb{F}[x] \oplus \mathbb{F}[y] \\ & & & \uparrow \\ & & & \mathbb{F}[x^{q-1}] \oplus \mathbb{F}[y^{q-1}] \end{array}$$

where $\mathbb{F}[x^{q-1}] = \mathcal{D}^*(1)_1$ and $\mathbb{F}[y^{q-1}] = \mathcal{D}^*(1)_2$. Then certainly

$$x^{q-1}, \text{ resp. } y^{q-1} \in \mathbb{F}[x,\ y]/(xy)$$

are zero divisors, and generate $\mathcal{P}^*$-invariant principal ideals. In this case it is easy to explicitly verify that

$$\begin{array}{cccc} L: & \mathbb{F}[x,\ y]/(xy) & \hookrightarrow & \mathbb{F}[x] \oplus \mathbb{F}[y] \\ & \uparrow & & \uparrow \\ & \mathbb{F}[x^{q-1} + y^{q-1}] & \hookrightarrow & \mathbb{F}[x^{q-1}] \oplus \mathbb{F}[y^{q-1}] \end{array}$$

completes our diagram in the desired way, i.e., $\mathbb{F}[x^{q-1} + y^{q-1}]$ is a Dickson algebra sitting integrally inside $\mathbb{F}[x,\ y]/(xy)$. Note

$$-(x^{q-1} + y^{q-1})x + x^q = 0,$$

and

$$-(x^{q-1} + y^{q-1})y + y^q = 0,$$

are the respective integral relations for x and y over $\mathbb{F}[x^{q-1} + y^{q-1}]$. Note also that the common lift of the top (and here only) Dickson class is exactly the coefficient of the Δ-relation, i.e.

$$-(x^{q-1} + y^{q-1})\mathcal{P}^{\Delta_0} + \mathcal{P}^{\Delta_1} = 0$$

on H^*, as it should be.

This example extends to the general case of Krull dimension 1. Specifically, continuing to use the above notation, we have:

Proposition 8.1.3: *Let* H^* *be a Noetherian unstable algebra of Krull dimension 1. Then* H^* *contains a fractal* $\mathcal{D}^*(1)^{q^s}$ *of the Dickson algebra of Krull dimension 1. Consider*

$$L : \mathrm{H}^*/\mathcal{N}il(\mathrm{H}^*) \hookrightarrow \bigoplus_{i=0}^{k} \mathrm{H}^*/\mathfrak{p}_i,$$

where we sum over all prime ideals $\mathfrak{p}_1, \dots, \mathfrak{p}_k$ *of height zero. Then*

$$\mathcal{D}^*(1)^{q^s} \cong \mathbb{F}[\mathbf{d}_{1,0}^{q^s}(1) + \cdots + \mathbf{d}_{1,0}^{q^s}(k)]$$

for a suitably large $s \in \mathbb{N}_0$*. If* H^* *is* $\mathcal{P}^*$*-inseparably closed then* $s = 0$.

Proof: By Lemma 8.1.1 and Lemma 7.4.1, it is no restriction to assume that H^* is reduced, and $\mathcal{P}^*$-inseparably closed. Let's consider, as above, the embedding

$$\begin{array}{lccc} L: & \mathrm{H}^* & \hookrightarrow & \bigoplus_{i=1}^{k} \mathrm{H}^*/\mathfrak{p}_i \\ & \cup\!\!\uparrow & & \cup\!\!\uparrow \\ & \mathbb{F}[\mathbf{d}_{1,0}^{q^s}(1)+\cdots+\mathbf{d}_{1,0}^{q^s}(k)] & \hookrightarrow & \bigoplus_{i=1}^{k} \mathbb{F}[\mathbf{d}_{1,0}^{q^s}(i)] \end{array}$$

where, with the preceding notations, we choose the integer s, as before, such that

$$s = \max\{s_j \mid j = 1, \ldots, k\}.$$

We have to show that the isomorphism (*in the category of* $\mathbb{F}$*-algebras*)

$$\begin{array}{rccc} \varphi: & \mathbb{F}[\mathbf{d}_{1,0}^{q^s}(1)+\cdots+\mathbf{d}_{1,0}^{q^s}(k)] & \xrightarrow{\sim} & \mathbb{F}[\mathbf{d}_{1,0}^{q^s}] = \mathcal{D}^*(1)^{q^s} \\ & \mathbf{d}_{1,0}^{q^s}(1)+\cdots+\mathbf{d}_{1,0}^{q^s}(k) & \longmapsto & \mathbf{d}_{1,0}^{q^s} \end{array}$$

is an isomorphism in the category of unstable algebras, i.e., we have to show that

$$\mathcal{P}^r\left(\varphi\left(\mathbf{d}_{1,0}^{q^s}(1)+\cdots+\mathbf{d}_{1,0}^{q^s}(k)\right)\right) = \varphi\left(\mathcal{P}^r\left(\mathbf{d}_{1,0}^{q^s}(1)+\cdots+\mathbf{d}_{1,0}^{q^s}(k)\right)\right)$$

for any $r \geq 0$. We do so by induction on r. For $r = 0$ there is nothing to show. So let $r > 0$. Then we have[5]

$$\begin{aligned} &\mathcal{P}^r\left(\mathbf{d}_{1,0}^{q^s}(1)+\cdots+\mathbf{d}_{1,0}^{q^s}(k)\right) \\ &= \mathcal{P}^r\left(\mathbf{d}_{1,0}^{q^s}(1)\right)+\cdots+\mathcal{P}^r\left(\mathbf{d}_{1,0}^{q^s}(k)\right) \\ &= \left(\mathcal{P}^{\frac{r}{q^s}}(\mathbf{d}_{1,0}(1))\right)^{q^s}+\cdots+\left(\mathcal{P}^{\frac{r}{q^s}}(\mathbf{d}_{1,0}(k))\right)^{q^s} \\ &= \left(\mathcal{P}^{\frac{r}{q^s}-1}\left(\mathbf{d}_{1,0}(1)\right)\mathbf{d}_{1,0}(1)+\cdots+\mathcal{P}^{\frac{r}{q^s}-1}\left(\mathbf{d}_{1,0}(k)\right)\mathbf{d}_{1,0}(k)\right)^{q^s} \\ &= \left(\mathcal{P}^{r-q^s}\left(\mathbf{d}_{1,0}^{q^s}(1)\right)\mathbf{d}_{1,0}^{q^s}(1)+\cdots+\mathcal{P}^{r-q^s}\left(\mathbf{d}_{1,0}^{q^s}(k)\right)\mathbf{d}_{1,0}^{q^s}(k)\right) \end{aligned}$$

where we made use of Proposition A.2.1 in the appendix. So, what's left to show is, that for all $r > 0$

$$\begin{aligned} &\mathcal{P}^r(\mathbf{d}_{1,0}^{q^s}(1))\mathbf{d}_{1,0}^{q^s}(1)+\cdots+\mathcal{P}^r(\mathbf{d}_{1,0}^{q^s}(k))\mathbf{d}_{1,0}^{q^s}(k) = \\ &\qquad \mathcal{P}^r\left(\mathbf{d}_{1,0}^{q^s}(1)+\cdots+\mathbf{d}_{1,0}^{q^s}(k)\right)\left(\mathbf{d}_{1,0}^{q^s}(1)+\cdots+\mathbf{d}_{1,0}^{q^s}(k)\right). \end{aligned}$$

Expand the right hand side to get

$$\mathcal{P}^r\left(\mathbf{d}_{1,0}^{q^s}(1)+\cdots+\mathbf{d}_{1,0}^{q^s}(k)\right)\cdot\left(\mathbf{d}_{1,0}^{q^s}(1)+\cdots+\mathbf{d}_{1,0}^{q^s}(k)\right)$$

---

[5] Recall the convention that $\mathcal{P}^i \equiv 0$ if $i \notin \mathbb{N}_0$.

$$= \left(\sum_{i=1}^{k} \mathcal{P}^r \left(\mathbf{d}_{1,0}^{q^s}(i) \right) \mathbf{d}_{1,0}^{q^s}(i) \right)$$
$$+ \left(\sum_{i=1}^{k} \mathcal{P}^r \left(\mathbf{d}_{1,0}^{q^s}(i) \right) \left(\mathbf{d}_{1,0}^{q^s}(1) + \cdots + \widehat{\mathbf{d}_{1,0}^{q^s}(i)} + \cdots + \mathbf{d}_{1,0}^{q^s}(k) \right) \right)$$
$$= \left(\sum_{i=1}^{k} \mathcal{P}^r \left(\mathbf{d}_{1,0}^{q^s}(i) \right) \mathbf{d}_{1,0}^{q^s}(i) \right) + \left(\sum_{i=1}^{k} \sum_{j=1, j \neq i}^{k} \mathcal{P}^r \left(\mathbf{d}_{1,0}^{q^s}(i) \right) \left(\mathbf{d}_{1,0}^{q^s}(j) \right) \right)$$
$$= \left(\sum_{i=1}^{k} \mathcal{P}^r \left(\mathbf{d}_{1,0}^{q^s}(i) \right) \mathbf{d}_{1,0}^{q^s}(i) \right) + 0$$

where the last equation follows from Lemma 8.1.2, and

$$\begin{aligned} &\mathcal{P}^r \left(\mathbf{d}_{1,0}^{q^s}(i) \right) \left(\mathbf{d}_{1,0}^{q^s}(j) \right) \\ &\quad \in \left(\mathfrak{p}_1 \cap \cdots \cap \widehat{\mathfrak{p}_i} \cap \cdots \cap \mathfrak{p}_k \right) \cdot \left(\mathfrak{p}_1 \cap \cdots \cap \widehat{\mathfrak{p}_j} \cap \cdots \cap \mathfrak{p}_k \right) \\ &\quad \subseteq \left(\mathfrak{p}_1 \cap \cdots \cap \widehat{\mathfrak{p}_i} \cap \cdots \cap \mathfrak{p}_k \right) \cap \left(\mathfrak{p}_1 \cap \cdots \cap \widehat{\mathfrak{p}_j} \cap \cdots \cap \mathfrak{p}_k \right) \\ &\quad = \left(\mathfrak{p}_1 \cap \cdots \cap \mathfrak{p}_k \right), \quad \text{for } i \neq j \\ &\quad = (0), \end{aligned}$$

because the nil radical is zero by assumption. Hence, putting this together gives

$$\begin{aligned} &\varphi \left(\mathcal{P}^r \left(\mathbf{d}_{1,0}^{q^s}(1) + \cdots + \mathbf{d}_{1,0}^{q^s}(k) \right) \right) \\ &= \varphi \left(\mathcal{P}^{r-q^s} \left(\mathbf{d}_{1,0}^{q^s}(1) \right) \mathbf{d}_{1,0}^{q^s}(1) + \cdots + \mathcal{P}^{r-q^s} \left(\mathbf{d}_{1,0}^{q^s}(k) \right) \mathbf{d}_{1,0}^{q^s}(k) \right) \\ &= \varphi \left(\mathcal{P}^{r-q^s} \left(\mathbf{d}_{1,0}^{q^s}(1) + \cdots + \mathbf{d}_{1,0}^{q^s}(k) \right) \left(\mathbf{d}_{1,0}^{q^s}(1) + \cdots + \mathbf{d}_{1,0}^{q^s}(k) \right) \right) \\ &= \varphi \left(\mathcal{P}^{r-q^s} \left(\mathbf{d}_{1,0}^{q^s}(1) + \cdots + \mathbf{d}_{1,0}^{q^s}(k) \right) \right) \varphi \left(\mathbf{d}_{1,0}^{q^s}(1) + \cdots + \mathbf{d}_{1,0}^{q^s}(k) \right) \\ &= \mathcal{P}^{r-q^s} \left(\varphi \left(\mathbf{d}_{1,0}^{q^s}(1) + \cdots + \mathbf{d}_{1,0}^{q^s}(k) \right) \right) \varphi \left(\mathbf{d}_{1,0}^{q^s}(1) + \cdots + \mathbf{d}_{1,0}^{q^s}(k) \right) \\ &= \mathcal{P}^{r-q^s} \left(\mathbf{d}_{1,0}^{q^s} \right) \mathbf{d}_{1,0}^{q^s} \\ &= \mathcal{P}^r \left(\mathbf{d}_{1,0}^{q^s} \right) \\ &= \mathcal{P}^r \left(\varphi \left(\mathbf{d}_{1,0}^{q^s}(1) + \cdots + \mathbf{d}_{1,0}^{q^s}(k) \right) \right). \end{aligned}$$

So, we have an isomorphism in the category of unstable algebras

$$\mathbb{F}[\mathbf{d}_{1,0}^{q^s}(1) + \cdots + \mathbf{d}_{1,0}^{q^s}(k)] \cong \mathcal{D}^*(1)^{q^s} \subset \mathrm{H}^*.$$

Since H^* is $\mathcal{P}^*$-inseparably closed, it also contains the $\mathcal{P}^*$-inseparable closure of $\mathcal{D}^*(1)^{q^s}$, i.e.,

$$\mathcal{D}^*(1) = \sqrt[\mathcal{P}^*]{\mathcal{D}^*(1)^{q^s}} \subset \sqrt[\mathcal{P}^*]{\mathrm{H}^*} = \mathrm{H}^*.$$

By construction this extension is also integral. •

One might be tempted to try to generalize Proposition 8.1.3 by taking just the sum of the *bottom* Dickson classes $\mathbf{d}^{q^s}_{n,n-1}(i)$, for $i = 1, \ldots, k$, in H^* and its appropriate Steenrod powers to produce a Dickson algebra inside H^*. This won't work as the following example shows.

EXAMPLE 2: Let x_1, x_2, x_3 be forms of degree one, $\mathbb{F} = \mathbb{F}_2$ the field with two elements and let $\mathrm{H}^* = \mathbb{F}[x_1, x_2, x_3]/(x_1x_2x_3)$. Then we have

$$L : \mathbb{F}[x_1, x_2, x_3]/(x_1x_2x_3) \hookrightarrow \mathbb{F}[x_2, x_3] \oplus \mathbb{F}[x_1, x_3] \oplus \mathbb{F}[x_1, x_2] \hookleftarrow \mathcal{D}^*(2)_1 \oplus \mathcal{D}^*(2)_2 \oplus \mathcal{D}^*(2)_3$$

where for pairwise distinct i, j, k we have

$$\mathbf{d}_{2,0}(i) = x_j^2x_k + x_jx_k^2$$
$$\mathbf{d}_{2,1}(i) = x_j^2 + x_k^2 + x_jx_k.$$

The sum of the bottom Dickson classes

$$\begin{aligned} D_1 &:= \mathbf{d}_{2,1}(1) + \mathbf{d}_{2,1}(2) + \mathbf{d}_{2,1}(3) \\ &= x_1x_2 + x_1x_3 + x_2x_3 \\ &\in \mathrm{H}^* \end{aligned}$$

maps under $\mathcal{P}^1$ to the sum of the top Dickson classes, as it should[6] in a proper Dickson algebra of Krull dimension 2 over the field with two elements, viz,

$$\begin{aligned} \mathcal{P}^1(D_1) &= x_1^2x_2 + x_1x_2^2 + x_1^2x_3 + x_1x_3^2 + x_2^2x_3 + x_2x_3^2 \\ &= \mathbf{d}_{2,0}(1) + \mathbf{d}_{2,0}(2) + \mathbf{d}_{2,0}(3) \\ &= D_0. \end{aligned}$$

The two elements D_0, D_1, so defined, even form a polynomial subalgebra in H^*, because they are algebraically independent[7] as the short calculation of the determinant of the generalized Jacobian shows:

$$\det \begin{bmatrix} \mathcal{P}^{\Delta_0}(D_0) & \mathcal{P}^{\Delta_0}(D_1) \\ \mathcal{P}^{\Delta_1}(D_0) & \mathcal{P}^{\Delta_1}(D_1) \end{bmatrix} = \det \begin{bmatrix} D_0 & 0 \\ 0 & D_0 \end{bmatrix} = D_0^2,$$

which is not a zero divisor in H^*. However $\mathbb{F}[D_0, D_1] \subseteq \mathrm{H}^*$ is *not* closed under the action of the Steenrod algebra. Consider e.g.

$$\mathcal{P}^2(D_0) = x_1^4x_2 + x_1x_2^4 + x_1^4x_3 + x_1x_3^4 + x_2^4x_3 + x_2x_3^4$$

which is not expressible[8] as a polynomial in D_0 and D_1.

[6] See Proposition A.2.1 in the Appendix.

[7] Compare Theorem A.4.1 in the Appendix.

[8] Note that the only homogeneous polynomial of degree 5 in $\mathbb{F}[D_0, D_1]$ is D_0D_1.

We come to one of the main results of this chapter. Consider a $\mathcal{P}^*$-inseparably closed reduced algebra H^*. As above we have the embeddings

$$\begin{array}{rcl} L : \mathrm{H}^* & \hookrightarrow & \bigoplus_{i=1}^{k} \mathrm{H}^*/\mathfrak{p}_i \\ & & \uparrow \\ & & \bigoplus_{i=1}^{k} \mathcal{D}^*(n)_i^{q^s}, \end{array}$$

where, we again choose

$$s := \max\{s_j \mid j = 1, \dots, k\}.$$

Denote by $\mathbf{d}_{n,0}^{q^s}(i) \in \mathcal{D}^*(n)_i^{q^s}$ the q^s-th fractal of the top Dickson class in the i-th summand. Define

$$\mathbf{t} := \mathbf{d}_{n,0}^{q^s}(1) + \cdots + \mathbf{d}_{n,0}^{q^s}(k) \in \mathrm{H}^*.$$

At this point we need a technical lemma.[9]

LEMMA 8.1.4: *Let H^* be a $\mathcal{P}^*$-inseparably closed reduced Noetherian unstable algebra, $\mathrm{H}^* \neq \mathbb{F}$, we have*

(1) *$\mathbf{t}$ is a non-zero divisor in H^*.*

(2) *The ideal $I := L^{-1}\left(\left((\mathbf{d}_{n,0}^{q^s}(1), \dots, \mathbf{d}_{n,0}^{q^s}(k))\right)\right) \subset \mathrm{H}^*$ is a $\mathcal{P}^*$-invariant ideal of height 1.*

PROOF: We take the statements in order.

AD (1) : Take an element $h \in \mathrm{H}^*$ such that

$$h\mathbf{t} = 0.$$

Projecting onto $\mathrm{H}^*/\mathfrak{p}_i$ gives

$$0 = \mathrm{pr}_i(h\mathbf{t}) = \mathrm{pr}_i(h)\mathrm{pr}_i(\mathbf{t}) = \mathrm{pr}_i(h)\mathbf{d}_{n,0}^{q^s}(i)$$

for every $i = 1, \dots, k$. Since the fractal of the Dickson class is not zero in $\mathrm{H}^*/\mathfrak{p}_i$ and $\mathrm{H}^*/\mathfrak{p}_i$ is an integral domain we deduce that

$$\mathrm{pr}_i(h) = 0,$$

i.e., $h \in \mathfrak{p}_i$ for every i. Hence

$$h \in \mathfrak{p}_1 \cap \cdots \cap \mathfrak{p}_k = (0).$$

So, h is zero.

---

[9] Note that the following statement only makes sense if there is a non-zero divisor in H^*. However, if H^* consists of zero divisors, and is, by our assumption reduced, then (0) being the intersection of its associated prime ideals and not having embedded components, must be the maximal ideal itself, i.e., H^* was $\mathbb{F}$, a trivial situation.

AD (2) : The ideal I is the preimage of a $\mathcal{P}^*$-invariant ideal, and therefore itself $\mathcal{P}^*$-invariant by Lemma 2.1 in [26]. Since $\mathbf{t} \in I$ and $\mathbf{t}$ is not a zero divisor by part (1), the ideal I has height[10] at least one. Moreover, as ideals in H^* we have[11]

$$\begin{aligned} I := & L^{-1}\left(\left((\mathbf{d}_{n,0}^{q^s}(1), \dots, \mathbf{d}_{n,0}^{q^s}(k))\right)\right) \\ \subseteq & L^{-1}\left(\left((\mathfrak{p}_k/\mathfrak{p}_1, \dots, \mathfrak{p}_k/\mathfrak{p}_{k-1}, \mathbf{d}_{n,0}^{q^s}(k))\right)\right) \\ \subseteq & \left(\mathfrak{p}_k, \mathbf{d}_{n,0}^{q^s}(k)\right), \end{aligned}$$

where the last statement follows from Lemma 8.1.2. By construction, $\mathfrak{p}_k \subset \mathrm{H}^*$ is a prime ideal of height 0 not containing $\mathbf{d}_{n,0}^{q^s}(k)$. Therefore the ideal

$$\left(\mathfrak{p}_k, \mathbf{d}_{n,0}^{q^s}(k)\right)$$

has height 1, and hence I as height at most 1. Putting the two things together gives that I has exactly height 1. •

We come to the second important result of this chapter.

THEOREM 8.1.5 (Big Imbedding Theorem): *Let H^* be a Noetherian unstable algebra over the Steenrod algebra. Then there is a fractal of the Dickson algebra, $\mathcal{D}^*(n)^{q^r}$, integrally in H^*. If H^* is $\mathcal{P}^*$-inseparably closed then $r = 1$,*

PROOF: By Lemma 8.1.1 and Lemma 7.4.1 we can, without loss of generality, assume that H^* has no nilpotent elements and is $\mathcal{P}^*$-inseparably closed. We proceed by induction on n. The case $n = 1$ is the contents of Proposition 8.1.3, so assume $n > 1$. We have

$$\begin{array}{ccc} L : \mathrm{H}^* & \hookrightarrow & \bigoplus_{i=1}^{k} \mathrm{H}^*/\mathfrak{p}_i \\ & & \cup \\ & & \bigoplus_{i=1}^{k} \mathcal{D}^*(n)_i^{q^s} \end{array}$$

and

$$\mathbf{d}_{n,0}^{q^s}(j) \in \mathfrak{p}_i \ \forall\ i \neq j.$$

Consider the element

$$\mathbf{d} := \left(\mathbf{d}_{n,0}^{q^s}(1), \dots, \mathbf{d}_{n,0}^{q^s}(k)\right) \in \bigoplus_{i=1}^{k} \mathrm{H}^*/\mathfrak{p}_i.$$

---

[10] Recall that the height of an arbitrary ideal I is defined to be the minimal height of a prime ideal containing I.

[11] I agree, this looks awful: $L^{-1}\left(\left((\mathbf{d}_{n,0}^{q^s}(1), \dots, \mathbf{d}_{n,0}^{q^s}(k))\right)\right)$ is the inverse image of the principal ideal $\left((\mathbf{d}_{n,0}^{q^s}(1), \dots, \mathbf{d}_{n,0}^{q^s}(k))\right)$ generated by the k-tupel $(\mathbf{d}_{n,0}^{q^s}(1), \dots, \mathbf{d}_{n,0}^{q^s}(k)) \in \oplus \mathrm{H}^*/\mathfrak{p}_i$.

By construction this element is a Thom class, i.e., it generates a height one $\mathcal{P}^*$-invariant principal ideal. The preimage under L of the ideal generated by $\mathbf{d}$ is the ideal I in Lemma 8.1.4. Hence we are in the lucky position that we have found a $\mathcal{P}^*$-invariant ideal $I \subset \mathrm{H}^*$ of height 1, mapping via L to the principal $(\mathbf{d}) \subset \bigoplus_{i=1}^{k} \mathrm{H}^*/\mathfrak{p}_i$, so we get a commutative diagram

$$\begin{array}{cccc} L: & \mathrm{H}^* & \hookrightarrow & \bigoplus_{i=1}^{k} \mathrm{H}^*/\mathfrak{p}_i \\ & \Big\downarrow \mathrm{pr} & © & \Big\downarrow \mathrm{pr} \\ \bar{L}: & \mathrm{H}^*/I & \hookrightarrow & \bigoplus_{i=1}^{k} \mathrm{H}^*/(\mathfrak{p}_i, \mathbf{d}_{n,0}^{q^s}(i)). \end{array}$$

By the fractal property (compare Proposition A.3.2) of the Dickson algebra the projection maps the pair

$$\bigoplus_{i=1}^{k} \mathcal{D}^*(n)_i^{q^s} \subseteq \bigoplus_{i=1}^{k} \mathrm{H}^*/\mathfrak{p}_i$$

onto the pair

$$\bigoplus_{i=1}^{k} \mathcal{D}^*(n-1)_i^{q^{s+1}} \subseteq \bigoplus_{i=1}^{k} \mathrm{H}^*/(\mathfrak{p}_i, \mathbf{d}_{n,0}^{q^s}(i)).$$

Since H^*/I has Krull dimension one less than H^*, we may apply the induction hypothesis, and conclude that there is a fractal of the Dickson algebra integrally inside H^*/I, say

$$\mathcal{D}^*(n-1)^{q^t} = \mathbb{F}[\mathbf{d}_{n-1,0}^{q^t}, \ldots, \mathbf{d}_{n-1,n-2}^{q^t}] \subseteq \mathrm{H}^*/I.$$

Since $\mathfrak{p}_i/I \subset \mathrm{H}^*/I$ has height zero this fractal is integrally contained in the quotient

$$\mathcal{D}^*(n-1)^{q^t} \subseteq \mathrm{H}^*/(\mathfrak{p}_i, I)$$

for every $i = 1, \ldots, k$. On the other hand, the fractals of the Dickson algebra

$$\mathcal{D}^*(n-1)_i^{q^{s+1}} \subseteq \mathrm{H}^*/(\mathfrak{p}_i, \mathbf{d}_{n,0}^{q^s}(i)),$$

for $i = 1, \ldots, k$ have also a common lift to

$$\mathcal{D}^*(n-1)^{q^{s+1}} \subseteq \mathrm{H}^*/I,$$

again by induction. Set $r := \max\{s+1, t\}$. Then we get, by the Uniqueness Theorem 7.3.3, that both fractals raised to the $q^{r-(s+1)}$, resp. q^{r-t}-th, power must be equal: Hence we have the following situation

$$\begin{array}{ccccc}
\mathrm{H}^* & \underset{L}{\hookrightarrow} & \bigoplus_{i=1}^{k} \mathrm{H}^*/\mathfrak{p}_i & \overset{\mathrm{pr}_i}{\longrightarrow} & \mathrm{H}^*/\mathfrak{p}_i \\
\downarrow \mathrm{pr} & \copyright & \downarrow \mathrm{pr} & \copyright & \downarrow \mathrm{pr} \\
\mathrm{H}^*/(I^{q^{r-s}}) & \underset{\bar{L}}{\hookrightarrow} & \bigoplus_{i=1}^{k} \mathrm{H}^*/(\mathfrak{p}_i, \mathbf{d}^{q^r}_{n,0}(i)) & \overset{\mathrm{pr}_i}{\longrightarrow} & \mathrm{H}^*/(\mathfrak{p}_i, \mathbf{d}^{q^r}_{n,0}(i)) \\
\uparrow & & \uparrow & & \uparrow \\
\mathcal{D}^*(n-1)^{q^r} & \hookrightarrow & \bigoplus_{i=1}^{k} \mathcal{D}^*(n-1)^{q^r}_i & \longrightarrow & \mathcal{D}^*(n-1)^{q^r}_i .
\end{array}$$

Again[12], by Theorem 7.3.3, the Dickson fractal in $\mathrm{H}^*/(\mathfrak{p}_i, \mathbf{d}^{q^r}_{n,0}(i))$ is equal to the fractal of the Dickson algebra, $\mathcal{D}^*(n-1)^{q^r}_i$ coming from $\mathrm{H}^*/\mathfrak{p}_i$ by projecting onto $\mathrm{H}^*/(\mathfrak{p}_i, I)$. So we arrive at a diagram as follows

$$\begin{array}{ccccc}
 & & \bigoplus_{i=1}^{k} \mathcal{D}^*(n)^{q^{r-1}}_i & \longrightarrow & \mathcal{D}^*(n)^{q^{r-1}}_i \\
 & & \downarrow & & \downarrow \\
\mathrm{H}^* & \underset{L}{\hookrightarrow} & \bigoplus_{i=1}^{k} \mathrm{H}^*/\mathfrak{p}_i & \overset{\mathrm{pr}_i}{\longrightarrow} & \mathrm{H}^*/\mathfrak{p}_i \\
\downarrow \mathrm{pr} & \copyright & \downarrow \mathrm{pr} & \copyright & \downarrow \mathrm{pr} \\
\mathrm{H}^*/(I^{q^{r-s}}) & \underset{\bar{L}}{\hookrightarrow} & \bigoplus_{i=1}^{k} \mathrm{H}^*/(\mathfrak{p}_i, \mathbf{d}^{q^r}_{n,0}(i)) & \overset{\mathrm{pr}_i}{\longrightarrow} & \mathrm{H}^*/(\mathfrak{p}_i, \mathbf{d}^{q^r}_{n,0}(i)) \\
\uparrow & & \uparrow & & \uparrow \\
\mathcal{D}^*(n-1)^{q^r} & \underset{\text{diagonally}}{\hookrightarrow} & \bigoplus_{i=1}^{k} \mathcal{D}^*(n-1)^{q^r}_i & \longrightarrow & \mathcal{D}^*(n-1)^{q^r}_i .
\end{array}$$

We look at the appropriate fractals of the bottom Dickson classes involved:

$$\begin{array}{ccccc}
\mathbf{d}^{q^r}_{n-1,n-2} + (I^{q^{r-s}}) & & \left(\mathbf{d}^{q^{r-1}}_{n,n-1}(1), \ldots, \mathbf{d}^{q^{r-1}}_{n,n-1}(k)\right) & \overset{\mathrm{pr}_i}{\longmapsto} & \mathbf{d}^{q^{r-1}}_{n,n-1}(i) \\
\downarrow \mathrm{pr} & & \downarrow & & \downarrow \mathrm{pr} \\
\mathbf{d}^{q^r}_{n-1,n-2} & \overset{\bar{L}}{\longmapsto} & \left(\mathbf{d}^{q^r}_{n-1,n-2}(1), \ldots, \mathbf{d}^{q^r}_{n-1,n-2}(k)\right) & \overset{\mathrm{pr}_i}{\longmapsto} & \mathbf{d}^{q^r}_{n-1,n-2}(i).
\end{array}$$

The composition maps $\mathrm{pr}_i \circ L$ and $\mathrm{pr}_i \circ \bar{L}$ are surjective. Therefore the bottom Dickson class in the upper right corner, $\mathbf{d}^{q^{r-1}}_{n,n-1}(i) \in \mathrm{H}^*/\mathfrak{p}_i$ must have a preimage in H^* and, by commutativity of this diagram, in the preimage of the lower left bottom Dickson class

$$(\mathrm{pr}_i \circ L)^{-1}(\mathbf{d}^{q^{r-1}}_{n,n-1}(i)) \in \mathbf{d}^{q^r}_{n-1,n-2} + (I^{q^{r-s}}) = \mathrm{pr}^{-1}(\mathbf{d}^{q^r}_{n-1,n-2}) \subseteq \mathrm{H}^*.$$

[12] If $\mathrm{H}^*/(\mathfrak{p}_i, I)$ is not reduced, we can't apply Theorem 7.3.3 directly; in this case we need to observe that our fractals of the Dickson algebra are equal in the quotient obtained by dividing out the nil radical. Then, since in a Noetherian ring, the nil radical is annihilated by some large q-power, we get that the two fractals of the Dickson algebra become equal after taking suitably high q-th powers, i.e., take just r to be large enough in the following discussion.

Or, in other words, there is an element[13]

$$\mathbf{t}_{n,n-1}^{q^{r-1}} := \mathbf{d}_{n-1,n-2}^{q^r} = (\mathrm{pr}_i \circ L)^{-1}(\mathbf{d}_{n,n-1}^{q^{r-1}}(i)) \in \mathrm{H}^* \quad \forall\, i = 1, \ldots, k,$$

mapping via pr to $\mathbf{d}_{n-1,\ n-2}^{q^r}$ and via $\mathrm{pr}_i \circ L$ to $\mathbf{d}_{n,n-1}^{q^{r-1}}(i)$ for all $i = 1, \ldots, k$. Set[14]

$$\mathbf{t}_{n,i}^{q^{r-1}} := \mathcal{P}^{q^i q^{r-1}} \cdots \mathcal{P}^{q^{n-1} q^{r-1}}(\mathbf{t}_{n,n-1}^{q^{r-1}})$$

for $i = 0, \ldots, n-2$, We claim that

$$\begin{array}{rccc} \varphi: & \mathbb{F}[\mathbf{t}_{n,0}^{q^{r-1}}, \ldots, \mathbf{t}_{n,n-1}^{q^{r-1}}] & \longrightarrow & \mathcal{D}^*(n)^{q^{r-1}} \\ & \mathbf{t}_{n,i}^{q^{r-1}} & \longmapsto & \mathbf{d}_{n,i}^{q^{r-1}} \end{array}$$

is an isomorphism of unstable algebras over the Steenrod algebra. To prove this we must show that

$$\varphi\left(\mathcal{P}^k\left(\mathbf{t}_{n,n-1}^{q^{r-1}}\right)\right) = \mathcal{P}^k\left(\varphi\left(\mathbf{t}_{n,n-1}^{q^{r-1}}\right)\right)$$

for every $k \geq 0$. We proceed inductively. For $k = 0$ there is nothing to show. So, assume that $k > 0$. Then by construction, for $i = 0, \ldots, n-1$,

$$\begin{aligned}
\mathcal{P}^k&\left(\varphi\left(\mathbf{t}_{n,i}^{q^{r-1}}\right)\right)\\
&= \mathcal{P}^k\left(\mathbf{d}_{n,i}^{q^{r-1}}\right)\\
&= \left(\mathcal{P}^{\frac{k}{q^{r-1}}}\left(\mathbf{d}_{n,i}\right)\right)^{q^{r-1}}\\
&= \left(-\mathcal{P}^{\frac{k}{q^{r-1}}-q^{i-1}}\left(\mathbf{d}_{n,i-1}\right) + \left(\mathcal{P}^{\frac{k}{q^{r-1}}-q^{n-1}}\left(\mathbf{d}_{n,n-1}\right)\right)\mathbf{d}_{n,i}\right)^{q^{r-1}}\\
&= \left(-\mathcal{P}^{\frac{k}{q^{r-1}}-q^{i-1}}\left(\varphi(\mathbf{t}_{n,i-1})\right) + \left(\mathcal{P}^{\frac{k}{q^{r-1}}-q^{n-1}}\left(\varphi(\mathbf{t}_{n,n-1})\right)\right)\varphi(\mathbf{t}_{n,i})\right)^{q^{r-1}}\\
&= \left(-\varphi\left(\mathcal{P}^{\frac{k}{q^{r-1}}-q^{i-1}}(\mathbf{t}_{n,i-1})\right) + \left(\varphi\left(\mathcal{P}^{\frac{k}{q^{r-1}}-q^{n-1}}(\mathbf{t}_{n,n-1})\right)\right)\varphi(\mathbf{t}_{n,i})\right)^{q^{r-1}}\\
&= \varphi\left(-\mathcal{P}^{\frac{k}{q^{r-1}}-q^{i-1}}(\mathbf{t}_{n,i-1}) + \left(\mathcal{P}^{\frac{k}{q^{r-1}}-q^{n-1}}(\mathbf{t}_{n,n-1})\right)(\mathbf{t}_{n,i})\right)^{q^{r-1}}\\
&= \varphi\left(\mathcal{P}^{\frac{k}{q^{r-1}}}(\mathbf{t}_{n,i})\right)^{q^{r-1}}\\
&= \varphi\left(\mathcal{P}^k(\mathbf{t}_{n,i}^{q^{r-1}})\right),
\end{aligned}$$

---

[13] More precisely: comparing degrees leads to the only possible value, namely

$$L^{-1} \circ \mathrm{pr}_i^{-1}(\mathbf{d}_{n,n-1}^{q^{r-1}}(i)) = \mathbf{d}_{n-1,n-2}^{q^r} \in \mathrm{H}^*$$

for all possible $i = 1, \ldots, k$ (unless $n = 1$ which is already covered by Proposition 8.1.3).

[14] This definition is motivated by Corollary A.2.2, that tells us how to jump from one Dickson class to the next with an appropriate Steenrod power.

where we made use of the fact that

$$\begin{aligned}\mathcal{P}^k(\mathbf{t}_{n,i}^{q^{r-1}}) &= \left(\mathcal{P}^{\frac{k}{q^{r-1}}}(\mathbf{t}_{n,i})\right)^{q^{r-1}}\\ &= \left(-\mathcal{P}^{\frac{k}{q^{r-1}}-q^{i-1}}(\mathbf{t}_{n,i-1}) + \left(\mathcal{P}^{\frac{k}{q^{r-1}}-q^{n-1}}(\mathbf{t}_{n,n-1})\right)(\mathbf{t}_{n,i})\right)^{q^{r-1}} + h_j\end{aligned}$$

for some $h_j \in \mathfrak{p}_j$ by construction. Since this is true for every $j = 1, \ldots, k$, we have that

$$\mathcal{P}^k(\mathbf{t}_{n,i}^{q^{r-1}}) = \left(-\mathcal{P}^{\frac{k}{q^{r-1}}-q^{i-1}}(\mathbf{t}_{n,i-1}) + \left(\mathcal{P}^{\frac{k}{q^{r-1}}-q^{n-1}}(\mathbf{t}_{n,n-1})\right)(\mathbf{t}_{n,i})\right)^{q^{r-1}} + h$$

for some $h \in \mathfrak{p}_1 \cap \cdots \cap \mathfrak{p}_k = (0)$. (We could have also just observed that the action of the Steenrod algebra commutes with the embedding L, and, for the image under L, these recursion formulae are true, because $L(\mathbf{t}_{n,n-1}^{q^{r-1}})$ is a fractal of the bottom Dickson class.) So we have $\mathcal{D}^*(n)^{q^{r-1}} \cong \mathbb{F}[\mathbf{t}_{n,0}^{q^{r-1}}, \ldots, \mathbf{t}_{n,n-1}^{q^{r-1}}] \subset \mathrm{H}^*$. Since we assumed that H^* is $\mathcal{P}^*$-inseparably closed we get that all p-th roots are in H^*, hence the Dickson algebra itself is in H^*

$$\mathcal{D}^*(n) \cong \mathbb{F}[\mathbf{t}_{n,0}, \ldots, \mathbf{t}_{n,n-1}] \subset \mathrm{H}^*$$

because it is the $\mathcal{P}^*$-inseparable closure of its fractals. •

REMARK: Note that by Theorem 7.3.3 the Dickson algebra we just found in a $\mathcal{P}^*$-inseparably closed H^* is unique.

EXAMPLE 3: In Lemma 8.1.1 we showed that one can pull back a Dickson algebra inside $\mathrm{H}^*/\mathcal{N}il(\mathrm{H}^*)$ to H^* with a possible necessary correction: one has to perhaps take a *fractal* of the Dickson algebra. However, here are two examples were we have the Dickson algebra itself integrally in H^* even though the nil radical is not trivial. Let x, y have degree one. Then

(1) $\mathcal{D}^*(1) = \mathbb{F}[x^{q-1}] \hookrightarrow \mathbb{F}[x, y]/(y^q)$ and

(2) $\mathcal{D}^*(1) = \mathbb{F}[x^{q-1}] \hookrightarrow \mathbb{F}[x, y^p]/(y^{q^2})$.

Note that the second example is not even $\mathcal{P}^*$-inseparably closed.

We derive a corollary as in the case of integral domains.

COROLLARY 8.1.6: *Let H^* be a Noetherian unstable algebra of Krull dimension n. Then there exists a homogeneous system of parameters $h_0, \ldots, h_{n-1}$ such that the polynomial subalgebra it generates is an unstable algebra over the Steenrod algebra.*

PROOF: Take $h_i := \mathbf{d}_{n,i}^{q^s}$ for suitable large s. •

8.2 Consequences for the Module of Derivations

In this section we derive some consequences from the Big Imbedding Theorem (8.1.5) for the module of derivations $\Delta(\mathrm{H}^*)$ defined in Section 1.2.

Throughout this section H^* is a reduced unstable algebra of finite Krull dimension n.

We start with proving the converse of Corollary 7.3.2.

PROPOSITION 8.2.1: *Let* H^* *be a reduced unstable algebra of Krull dimension* n. *Then the following are equivalent:*

(1) H^* *is Noetherian.*

(2) H^* *is* Δ*-finite, has finitely many height zero prime ideals and contains integrally a fractal of the Dickson algebra* $\mathcal{D}^*(n)$.

PROOF: If H^* is Noetherian, then it is Δ-finite by Corollary 1.2.2. By the Imbedding Theorem 8.1.5 it contains integrally a fractal of the Dickson algebra, and, certainly has only finitely many height zero prime ideals.

Conversely, if H^* satisfies all three conditions, then Corollary 7.3.2 proves that H^* is Noetherian. •

We come back to the investigation of the H^*-module $\Delta(\mathrm{H}^*)$ of derivations generated by $\{\mathcal{P}^{\Delta_i} \mid i \geq 0\}$. See Section 1.2 and, in particular, also Example 1 in that Section.

PROPOSITION 8.2.2: *Let* H^* *be a reduced* $\mathcal{P}^*$*-inseparably closed unstable algebra of Krull dimension* n. *If* H^* *is Noetherian then the* H^**-module* $\Delta(\mathrm{H}^*)$ *of derivations is given by*

$$\Delta(\mathrm{H}^*) = \bigoplus_{i=0}^{n-1} \mathrm{H}^* \mathcal{P}^{\Delta_i}.$$

In particular it is a free module on n *generators.*

PROOF: By Proposition 7.3.1 the Δ-relation for H^* is

$$(-1)^n \mathbf{d}_{n,\,0} \mathcal{P}^{\Delta_0} + \cdots + (-1)\mathbf{d}_{n,\,n-1} \mathcal{P}^{\Delta_{n-1}} + \mathcal{P}^{\Delta_n} = 0.$$

Therefore

$$\mathcal{P}^{\Delta_n} \in \bigoplus_{i=0}^{n-1} \mathrm{H}^* \mathcal{P}^{\Delta_i}.$$

Applying Lemma 1.1.8 yields

$$\mathcal{P}^{\Delta_n + r} \in \bigoplus_{i=r}^{n+r-1} \mathrm{H}^* \mathcal{P}^{\Delta_i}$$

for every $r \in \mathbb{N}_0$. Hence inductively we get

$$\mathcal{P}^{\Delta_n + r} \in \bigoplus_{i=0}^{n-1} \mathrm{H}^* \mathcal{P}^{\Delta_i}$$

for all $r \in \mathbb{N}_0$, where freeness follows from minimality. •

The following examples illustrate that the conclusion of the preceding result may remain true under somewhat weaker assumptions.

EXAMPLE 1: Define

$$\mathrm{H}^* := \mathbb{F}[x_1, x_2, x_3, \ldots]/(x_1^p, x_2^p, x_3^p, \ldots).$$

Then H^* has Δ-relation

$$\mathcal{P}^{\Delta_0} = 0$$

and therefore the module of derivations $\Delta(\mathrm{H}^*) = 0$ is trivial and hence free. However, H^* is not Noetherian.

EXAMPLE 2: Have a look at

$$\mathrm{H}^* := \mathbb{F}[x_1, x_2, x_3, \ldots]/(x_i x_j, \forall\, i \neq j),$$

where all the generators, x_i, $i \geq 1$, have degree one. This time H^* is reduced $\mathcal{P}^*$-inseparably closed of Krull dimension 1. The Δ-relation has length 1

$$-x_1^{q-1}\mathcal{P}^{\Delta_0} + \mathcal{P}^{\Delta_1} = 0$$

and therefore the module of derivations $\Delta(\mathrm{H}^*) = \mathrm{H}^*\mathcal{P}^{\Delta_0}$ is free on one generator. However, H^* fails to be Noetherian, because there are infinitely many height zero prime ideals, namely

$$\mathfrak{p}_i := (x_1, \ldots, x_{i-1}, \widehat{x_i}, x_{i+1}, \ldots) \quad \forall\, i \in \mathbb{N}.$$

If H^* is not $\mathcal{P}^*$-inseparably closed the situation is a bit more delicate. The first example that comes into mind (recall Example 2 of Section 6.1) is

$$\mathrm{H}^* := \mathbb{F}[x_1^{p^s}, x_2^{p^s}, x_3^{p^s}, \ldots]$$

has Δ-length zero, and hence trivial H^*-module $\Delta(\mathrm{H}^*)$, but it fails to be Noetherian. However, we can generalize the preceding result in the following way:

PROPOSITION 8.2.3: *Let H^* be a reduced unstable algebra of Krull dimension n. If H^* is Noetherian then the H^*-module $\Delta(\mathrm{H}^*)$ of derivations has at most $n + r$ generators, where*

$$\mathcal{D}^*(n)^{q^r} \hookrightarrow \mathrm{H}^*$$

is a fractal of the Dickson algebra, imbedding integrally into H^*.

PROOF: The Big Imbedding Theorem 8.1.5 guarantees the existence of a fractal of the Dickson algebra in H^*, such that

$$\mathcal{D}^*(n)^{q^r} \hookrightarrow \mathrm{H}^*$$

is integral. By Corollary 7.3.2 we have

$$(-1)^n \mathbf{d}_{n,0}^{q^r}\mathcal{P}^{\Delta_r} + \cdots + (-1)\mathbf{d}_{n,n-1}^{q^r}\mathcal{P}^{\Delta_{n-1+r}} + \mathcal{P}^{\Delta_{n+r}} = 0$$

on H^*. Therefore, by successive use of Lemma 1.1.8, we get

$$\mathcal{P}^{\Delta_{n+r+k}} \in \sum_{i=0}^{n+r} \mathrm{H}^* \mathcal{P}^{\Delta_i}$$

for all $k \in \mathbb{N}_0$. •

8.3 Turkish Delights II

Obviously the Big Imbedding Theorem gives a Thom class explicitly. We are going to derive some results exemplifying why Thom classes are wonderful: they allow proofs by induction *within our category*.

Throughout this section we are dealing only with Noetherian unstable algebras.

PROPOSITION **8.3.1** (Turkish Delight 5): *In every Noetherian unstable algebra* H^* *there is a Thom class* $\mathbf{t}$.

PROOF: Since we have found a fractal of the Dickson algebra in H^*, the top Dickson class $\mathbf{d}_{n,0}^{q^s}$ serves as the desired Thom class, compare Theorem 8.1.5 and Proposition A.2.1 in the Appendix. •

PROPOSITION **8.3.2** (Turkish Delight 6): *The Thom class* $\mathbf{t} = \mathbf{d}_{n,0}^{q^s}$ *which we just found in* H^*, *is an element of every* $\mathcal{P}^*$*-invariant prime ideal of positive height.*

PROOF: Since

$$\mathcal{D}^*(n)^{q^s} \hookrightarrow \mathrm{H}^*$$

is integral, for every $\mathcal{P}^*$-invariant homogeneous prime ideal

$$\mathfrak{q} \subset \mathrm{H}^*$$

of positive height, also the contracted ideal

$$\mathfrak{q} \cap \mathcal{D}^*(n)^{q^s}$$

is $\mathcal{P}^*$-invariant homogeneous prime ideal of positive height. By [21] (see Theorem 11.4.6 [36] for a complete proof) the only $\mathcal{P}^*$-invariant prime ideals in $\mathcal{D}^*(n)^{q^s}$ are

$$(0) \subsetneq (\mathbf{d}_{n,0}^{q^s}) \subsetneq (\mathbf{d}_{n,0}^{q^s}, \mathbf{d}_{n,1}^{q^s}) \subsetneq \cdots \subsetneq (\mathbf{d}_{n,0}^{q^s}, \ldots, \mathbf{d}_{n,n-1}^{q^s}).$$

Hence

$$\mathbf{d}_{n,0}^{q^s} \in \mathfrak{q} \cap \mathcal{D}^*(n)^{q^s}$$

and therefore also

$$\mathbf{d}_{n,0}^{q^s} \in \mathfrak{q}$$

as claimed. •

So, we have a Thom class with nice properties. Since Thom classes generate (by definition) $\mathcal{P}^*$-invariant height one ideals they allow us to construct proofs by induction over the Krull dimension as we will illustrate with the next two results.

The first one reproves Turkish Delight 4 (Proposition 6.2.3).

PROPOSITION 8.3.3 (Turkish Delight 7): *Let $\mathfrak{p} \in \mathrm{H}^*$ a $\mathcal{P}^*$-invariant homogeneous prime ideal of height $i > 0$, then there exist a $\mathcal{P}^*$-invariant prime ideal $\mathfrak{q} \subset \mathfrak{p}$ of height $i - 1$.*

PROOF: By double induction on the Krull dimension of H^* and the height i of $\mathfrak{p}$.

Let $\dim(\mathrm{H}^*) = 1$. Then there is nothing to show since prime ideals of height zero are $\mathcal{P}^*$-invariant by a result of Peter S. Landweber, [20] (see Proposition 11.2.3 in [36] for a complete proof or Theorem 3.5 in [29] for a more general result). So, let $\dim(\mathrm{H}^*) > 1$. We induct on the height of $\mathfrak{p}$. If $i = 1$ there is nothing to show since any prime ideal of height zero is $\mathcal{P}^*$-invariant. Let $i > 1$. By Turkish Delight 6 (Proposition 8.3.2) there exists a Thom class $\mathbf{t} \in \mathfrak{p}$. Dividing out the principal ideal generated by this Thom class we get an unstable algebra $\mathrm{H}^*/(\mathbf{t})$ over the Steenrod algebra of Krull dimension one less. Hence by induction there exists a $\mathcal{P}^*$-invariant homogeneous prime ideal

$$\overline{\mathfrak{q}} \subsetneq \mathfrak{p}/(\mathbf{t}) \subsetneq \mathrm{H}^*/(\mathbf{t})$$

with

$$\begin{aligned} \mathcal{h}t(\overline{\mathfrak{q}}) &= \mathcal{h}t(\mathfrak{p}/(\mathbf{t})) - 1 \\ &= i - 2. \end{aligned}$$

Hence the preimage in H^*

$$\mathfrak{q} = (\overline{\mathfrak{q}}, \mathbf{t}) \subsetneq \mathfrak{p} \subset \mathrm{H}^*$$

is a $\mathcal{P}^*$-invariant homogeneous prime ideal of height $i - 1$. •

We are now ready to prove the main result of this section: a $\mathcal{P}^*$-invariant version of Krull's principal ideal theorem and its generalization, see Theorem 1.2.16 in [5].

THEOREM 8.3.4 (Turkish Delight 8): *Let $\mathfrak{p}_i \subset \mathrm{H}^*$ be a $\mathcal{P}^*$-invariant homogeneous prime ideal of height i, for $i = 1, \ldots, n$. Then there exist elements $h_1, \ldots, h_i \in \mathrm{H}^*$ such that*

(1) *$h_1, \ldots, h_i \in \mathfrak{p}_i$,*
(2) *$\mathfrak{p}_i$ is an isolated prime ideal of $(h_1, \ldots, h_i)$, and*
(3) *$(h_1, \ldots, h_j) \subset \mathrm{H}^*$ are $\mathcal{P}^*$-invariant ideals of height j, $j = 1, \ldots, i$.*

PROOF: We use induction on the height i. The statement for $i = 1$ is the contents of Turkish Delight 6 (Proposition 8.3.2). So, suppose $i > 1$ and assume by induction that for any $\mathcal{P}^*$-invariant homogeneous prime ideal of height $< i$ the result is established, and let $\mathfrak{p}$ be a $\mathcal{P}^*$-invariant homogeneous prime ideal of height i. Then by Turkish Delight 7 (Proposition 8.3.3) there exists a $\mathcal{P}^*$-invariant homogeneous prime ideal $\mathfrak{q} \subsetneq \mathfrak{p}$ of height $i-1$. By the induction hypothesis there exist elements $h_1, \ldots, h_{i-1}$ such that

(1) $h_1, \ldots, h_{i-1} \in \mathfrak{q}$,

(2) $\mathfrak{q}$ is an isolated prime ideal of $(h_1, \ldots, h_{i-1})$, and

(3) The ideals $(h_1, \ldots, h_j) \subset \mathrm{H}^*$ are $\mathcal{P}^*$-invariant ideals of height j, $j = 1, \ldots, i-1$.

The quotient $\mathrm{H}^*/(h_1, \ldots, h_{i-1})$ is again an unstable algebra over the Steenrod algebra, and hence contains a Thom class $\mathbf{t}$ by Turkish Delight 5 (Proposition 8.3.1), which moreover is contained in $\overline{\mathfrak{p}} := \mathfrak{p}/(h_1, \ldots, h_{i-1})$ by Turkish Delight 6 (Proposition 8.3.2) (note that $\overline{\mathfrak{p}}$ has height one). If we choose a preimage $h_i \in \mathrm{pr}^{-1}(\mathbf{t})$ under the projection

$$\mathrm{pr} : \mathrm{H}^* \longrightarrow \mathrm{H}^*/(h_1, \ldots, h_{i-1})$$

then

$$(h_1, \ldots, h_i) \subseteq \mathfrak{p} \subset \mathrm{H}^*$$

is a $\mathcal{P}^*$-invariant ideal of height i and $\mathfrak{p}$ is an isolated prime ideal of it. •

8.4 The Reverse Landweber-Stong Conjecture

In the preceding section we have found a Thom class $\mathbf{t}$ in every unstable algebra over $\mathcal{P}^*$. In this section we prove that if the algebra in question does not consist of zero divisors, and is reduced, then our naturally given Thom class is a non-zero divisor.

Again H^* is assumed to be Noetherian and unstable.

PROPOSITION 8.4.1 (Turkish Delight 9): *Let $\mathcal{N}il(\mathrm{H}^*) = (0)$. If H^* contains a non-zero divisor, then it contains a non-zero divisor $\mathbf{t}$ which is a Thom class.*

PROOF: Let's go back a step and look how we constructed a Thom class in H^*. First we took the $\mathcal{P}^*$-inseparable closure $\sqrt[\mathcal{P}^*]{\mathrm{H}^*}$. If H^* contains a non-zero divisor then the Krull dimension n of H^* is positive. So by the Big Imbedding Theorem (8.1.5) there is a non-trivial Dickson algebra integrally inside the $\mathcal{P}^*$-inseparably closure, say

$$\mathcal{D}^*(n) \hookrightarrow \sqrt[\mathcal{P}^*]{\mathrm{H}^*},$$

which also has Krull dimension n. The top Dickson class is a Thom class by Turkish Delight 5 (Proposition 8.3.1), and, by Lemma 8.1.4, gives a non-zero divisor

$$\mathbf{d}_{n,0} \in \mathcal{D}^*(n) \hookrightarrow \sqrt[\mathcal{P}^*]{\mathrm{H}^*}.$$

Hence some q-th power of it is still a Thom class and still a non-zero divisor in H^*. •

This result does not remain true if we drop the condition on H^* that $\mathcal{N}il(\mathrm{H}^*) = (0)$ as the following example shows.

EXAMPLE 1: Let $\mathbb{F} = \mathbb{F}_2$ be the field with two elements, and let x, y, z be linear form. Consider the algebra

$$\mathrm{H}^* := \mathbb{F}[x, y, z]/(yz, z^2),$$

and note that its nil radical is

$$\mathcal{N}i\ell(\mathrm{H}^*) = (z) \neq (0).$$

Then H^* contains integrally a copy of the Dickson algebra $\mathcal{D}^*(2)$

$$\mathcal{D}^*(2) = \mathbb{F}[xy(x+y),\ x^2 + y^2 + xy] \hookrightarrow \mathrm{H}^*,$$

where $\mathbf{d}_{2,0} = xy(x+y)$ and $\mathbf{d}_{2,1} = x^2 + y^2 + xy$. Integrality follows from

$$\begin{aligned} 0 &= z^2, \\ 0 &= \mathbf{d}_{2,0}x + \mathbf{d}_{2,1}x^2 + x^4, \\ 0 &= \mathbf{d}_{2,0}y + \mathbf{d}_{2,1}y^2 + y^4. \end{aligned}$$

The top Dickson class is a Thom class, as Turkish Delight 5 (Proposition 8.3.1) predicts, but it is a zero divisor

$$\mathbf{d}_{2,0}z = xy(x+y)z = 0.$$

If Turkish Delight 9 (Proposition 8.4.1) would have been true without the additional assumption that the algebra be reduced, then it would be a routine chore to deduce by induction on the Krull dimension the reverse version of the Landweber-Stong conjecture. Recall some notions from the introduction:

A sequence $h_1, \ldots, h_k \in \mathrm{H}^*$ of elements of positive degree in H^* is called a **regular sequence** if

(1) $h_1 \in \mathrm{H}^*$ is not a zero divisor,
(2) $h_i \in \mathrm{H}^*/(h_1, \ldots, h_{i-1})$ is not a zero divisor $\forall\ i = 2, \ldots, k$.

Then define the **depth** (or **homological codimension**) of H^*, denoted $dp(\mathrm{H}^*)$, to be the length of the longest possible regular sequence in H^*. See [36] Chapter 6 for an introduction to the homological properties of graded connected commutative algebras.

The original **Landweber-Stong Conjecture** asserts that a ring of invariants $\mathbb{F}[V]^G$ has depth at least k, $dp\,(\mathrm{H}^*) \geq k$, if and only if the k *bottom* Dickson classes $\mathbf{d}_{n,n-1}, \ldots, \mathbf{d}_{n,n-k}$ form a regular sequence[15], see [22]. This conjecture was proven in 1996 by Dorra Bourguiba and Saïd Zarati, [6], using the classification of injective $\mathbb{F}[V]$-modules over $\mathcal{P}^*$ by J. Lannes and S. Zarati, see [23].

The **Reverse Landweber-Stong Conjecture**, [38], says that a ring of invariants $\mathbb{F}[V]^G$ has depth at least k if and only if the *top* k Dickson classes,

[15] Note that since we have a ring of invariants, of course, the Dickson algebra $\mathcal{D}^*(n) \hookrightarrow \mathbb{F}[V]^G$ is integrally contained in the ring of invariants.

$\mathbf{d}_{n,0}, \ldots, \mathbf{d}_{n,k-1}$ form a regular sequence. Since the Big Imbedding Theorem hands us a fractal of the Dickson algebra in any unstable algebra over the Steenrod algebra, we can reformulate the reverse Landweber-Stong conjecture in the following way:

THE REVERSE LANDWEBER-STONG CONJECTURE: Let H^* be a Noetherian unstable algebra over the Steenrod algebra. Then H^* has depth at least k if and only if high enough q-th powers of the k top Dickson classes, $\mathbf{d}^{q^s}_{n,0}, \ldots, \mathbf{d}^{q^s}_{n,k-1} \in H^*$ form a regular sequence.

This conjecture was one of the original motivations to study Thom classes, their existence and their properties. However, the above example shows, not only that $\mathcal{Nil}(H^*) = (0)$ is a necessary condition in Turkish Delight 9 (Proposition 8.4.1), it is also a counter example to the Reverse Landweber Stong Conjecture.

EXAMPLE 2: Recall that we have an integral extension of unstable Noetherian algebras

$$\mathcal{D}^*(2) = \mathbb{F}[xy(x+y),\ x^2+y^2+xy] \hookrightarrow H^* := \mathbb{F}[x,\ y,\ z]/(yz,\ z^2).$$

The big algebra H^* has depth 1, but the top Dickson class

$$\mathbf{d}_{2,0} = xy(x+y) \in H^*$$

is a zero divisor, as we have seen already. However, the bottom Dickson class

$$\mathbf{d}_{2,1} = x^2 + y^2 + xy \in H^*$$

is not a zero divisor as one can prove in the following way: let $h \in H^*$ such that

$$\left(x^2+y^2+xy\right) h = 0 \in (z) = \mathcal{Nil}(H^*) \subset H^*.$$

Hence $h = z\bar{h} \in (z)$, because (z) is a prime ideal and $\mathbf{d}_{2,1} \notin (z)$. Then

$$\begin{aligned} 0 &= \left(x^2+y^2+xy\right) h \\ &= \left(x^2+y^2+xy\right) z\bar{h} \\ &= x^2 z\bar{h} + (x+y)yz\bar{h} \\ &= x^2 z\bar{h} + 0 \\ &= x^2 z\bar{h} \end{aligned}$$

Since x^2 is not a zero divisor in H^* we conclude that

$$h = z\bar{h} = 0,$$

and that's all we wanted. •

We are almost at the end of our journey. What follows is an appendix with some technical proofs in all their detail. I want to say good-bye. It has been a long way and I hope you enjoyed it. And above all, I want to thank you for the company.

Appendix A
Technical Stuff

This appendix is full of unpleasant technical stuff.

A.1 Old and New Relations in the Steenrod Algebra

In this first section of the appendix we will prove the two technical lemmas whose proofs were omitted in Chapter 1. Recall from Chapter 1 that we made the convention that $\mathcal{P}^i = 0$ for $i \notin \mathbb{N}_0$.

We first prove Lemma 1.1.2, i.e.,

LEMMA A.1.1: *For all natural numbers $i \geq 1$ the following commutation rule holds*

$$\mathcal{P}^k \mathcal{P}^{\Delta_i} - \mathcal{P}^{\Delta_i} \mathcal{P}^k = \mathcal{P}^{\Delta_{i+1}} \mathcal{P}^{k-q^i}.$$

If $i = 0$ we have

$$\mathcal{P}^k \mathcal{P}^{\Delta_0} - \mathcal{P}^{\Delta_0} \mathcal{P}^k = k\mathcal{P}^k.$$

PROOF: As we already pointed out in Chapter 1 the second statement is clear, because $\mathcal{P}^{\Delta_0}$ is just multiplication with the degree of the operand.

For the first statement we translate the formula into Milnor's notation. We employ the notation $\mathcal{P} \hat{=} P$ to indicate that $\mathcal{P}$ and P are the same Steenrod operation written out in different notations; $\mathcal{P}$ for our standard notation, P for Milnor's. For example:

$$\mathcal{P}^k \hat{=} P^{(k,0,\dots)} \quad \text{and} \quad \mathcal{P}^{\Delta_i} \hat{=} P^{(0,\dots,0,1_i,0,\dots)},$$

where the index i at the 1 indicates that the 1 comes in the i-th position. Then the statement reads as follows:

$$P^{(k,0,\dots)} P^{(0,\dots,0,1_i,0,\dots)} - P^{(0,\dots,0,1_i,0,\dots)} P^{(k,0,\dots)} = P^{(0,\dots,0,1_{i+1},0,\dots)} P^{(k-q^i,0,\dots)}.$$

Following Milnor, let

$$R := (r_1,\ r_2, \ldots),$$

and

$$S := (s_1,\ s_2,\ \ldots),$$

and consider infinite matrices

$$\mathbf{X} = \begin{bmatrix} * & x_{01} & x_{02} & \cdots \\ x_{10} & x_{11} & x_{12} & \cdots \\ x_{20} & x_{21} & x_{22} & \cdots \\ \cdots & \cdots & \cdots & \cdots \end{bmatrix}$$

with entries $x_{ab} \in \mathbb{N}_0$ (and nothing in the 00-position) which satisfy the following system of equations

$$r_a = \sum_b q^b x_{ab}$$

$$s_b = \sum_a x_{ab}.$$

Then the product $P^R P^S$ is given by

$$P^R P^S = \sum_{\mathbf{X}} b(\mathbf{X}) P^{T(\mathbf{X})},$$

where the index sequence $T(\mathbf{X}) = (t_1,\ t_2, \ldots)$ is the sequence of diagonal sums

$$t_c = \sum_{a+b=c} x_{ab},$$

and the coefficient $b(\mathbf{X})$ is given by

$$b(\mathbf{X}) = \frac{\prod_c t_c!}{\prod_{a,\ b} x_{ab}!}.$$

Hence, for our product $P^{(k,0,\ldots)} P^{(0,\ldots,\ 0,1_i,0,\ldots)}$, we have to consider matrices $\mathbf{X}$ with

$$\sum_b q^b x_{ab} = \begin{cases} k & \text{if } a = 1 \\ 0 & \text{if } a > 1 \end{cases}$$

$$\sum_a x_{ab} = \begin{cases} 1 & \text{if } b = i \\ 0 & \text{if } b \leq 1,\ \neq i. \end{cases}$$

Therefore our matrices look like

$$\mathbf{X} = \begin{bmatrix} * & 0 & \cdots & 0 & x_{0i} & 0 & \cdots \\ x_{10} & 0 & \cdots & 0 & x_{1i} & 0 & \cdots \\ 0 & 0 & \cdots & 0 & 0 & 0 & \cdots \\ \cdots & \cdots & \cdots & \cdots & \cdots & \cdots & \cdots \end{bmatrix},$$

with

$$x_{0i} + x_{1i} = 1,$$
$$x_{10} + q^i x_{1i} = k.$$

This system has the following solutions

$$\mathbf{X}_1 = \begin{bmatrix} * & 0 & \cdots & 0 & 1_{0i} & 0 & \cdots \\ k & 0 & \cdots & 0 & 0 & 0 & \cdots \\ 0 & 0 & \cdots & 0 & \cdots & 0 & \cdots \\ \cdots & \cdots & \cdots & \cdots & \cdots & \cdots & \cdots \end{bmatrix},$$

and in addition, if $k \geq q^i$,

$$\mathbf{X}_2 = \begin{bmatrix} * & 0 & \cdots & 0 & 0 & 0 & \cdots \\ k - q^i & 0 & \cdots & 0 & 1_{1i} & 0 & \cdots \\ 0 & 0 & \cdots & 0 & 0 & 0 & \cdots \\ \cdots & \cdots & \cdots & \cdots & \cdots & \cdots & \cdots \end{bmatrix}$$

also. Hence we have

$$\begin{aligned}
&\mathcal{P}^k \mathcal{P}^{\Delta_i} \\
&\quad \hat{=}\, P^{(k,0,\ldots)} P^{(0,\ldots,0,1_i,0,\ldots)} \\
&\quad = \begin{cases} P^{(k,0,\ldots,0,1_i,0\ldots)} & \text{if } k < q^i \\ P^{(k,0,\ldots,0,1_i,0\ldots)} + P^{(k-q^i,0,\ldots,0,1_{i+1},0\ldots)} & \text{otherwise} \end{cases} \\
&\quad = \begin{cases} P^{(0,\ldots,0,1_i,0,\ldots)} P^{(k,0,\ldots)} & \text{if } k < q^i \\ P^{(0,\ldots,0,1_i,0,\ldots)} P^{(k,0,\ldots)} + P^{(0,\ldots,0,1_{i+1},0,\ldots)} P^{(k-q^i,0,\ldots)} & \text{otherwise} \end{cases} \\
&\quad \hat{=}\, \mathcal{P}^{\Delta_i} \mathcal{P}^k + \mathcal{P}^{\Delta_{i+1}} \mathcal{P}^{k-q^i},
\end{aligned}$$

as we claimed. •

Next comes the proof of Lemma 1.1.3.

LEMMA A.1.2: *Let $h \in \mathrm{H}^*$ be an unstable element of degree d. Then for any $r \geq 0$ and for any $s \geq 0$ there exists an element $\mathcal{M}_{r,d} \in \mathcal{P}^*$ such that*

$$\mathcal{P}^{\Delta_{r+s}}\left(\mathcal{M}_{r,d}(h)\right) = \left(\mathcal{P}^{\Delta_s}(h)\right)^{q^r},$$

where the notation $\mathcal{M}_{r,d}$ emphasizes the dependence on r and d.

PROOF: Set $d_0 := d$ and define inductively

$$\begin{aligned}
\mathcal{M}_{0,d} &:= \mathcal{P}^0 \\
\mathcal{M}_{1,d} &:= \mathcal{P}^{d_0-1} \\
\mathcal{M}_{r+1,d} &:= \mathcal{P}^{d_r-1}\mathcal{M}_{r,d}, \quad \text{where } d_r := \deg\left(\mathcal{M}_{r,d}(h)\right).
\end{aligned}$$

We verify the desired formula by induction on r. For $r = 0$ there is nothing to prove. So, let $r > 0$. Then, we have

$$\begin{aligned}
\mathcal{P}^{\Delta_{r+s}}\mathcal{M}_{r,d}(h) := \mathcal{P}^{\Delta_{r+s}}\mathcal{P}^{d_{r-1}-1}\mathcal{M}_{r-1,d}(h) \\
&\overset{(1)}{=} \left(\mathcal{P}^{d_{r-1}-1+q^{r+s-1}}\mathcal{P}^{\Delta_{r+s-1}} - \mathcal{P}^{\Delta_{r+s-1}}\mathcal{P}^{d_{r-1}-1+q^{r+s-1}}\right)\mathcal{M}_{r-1,d}(h) \\
&\overset{(2)}{=} \mathcal{P}^{d_{r-1}-1+q^{r+s-1}}\mathcal{P}^{\Delta_{r+s-1}}\mathcal{M}_{r-1,d}(h) - 0 \\
&\overset{(3)}{=} \mathcal{P}^{d_{r-1}-1+q^{r+s-1}}\left(\left(\mathcal{P}^{\Delta_s}(h)\right)^{q^{r-1}}\right) \\
&\overset{(4)}{=} \left(\left(\mathcal{P}^{\Delta_s(h)}\right)^{q^{r-1}}\right)^q \\
&= \left(\mathcal{P}^{\Delta_s(h)}\right)^{q^r},
\end{aligned}$$

where

(1) follows from Lemma A.1.1 with $i = r + s - 1$ and $k = d_{r-1} - 1 + q^{r+s-1}$,

(2) follows from the fact that

$$d_{r-1} - 1 + q^{r+s-1} \geq d_{r-1} = \deg\left(\mathcal{M}_{r-1,d}\right) + d,$$

(3) follows by induction, and

(4) by

$$\begin{aligned}
d_{r-1} - 1 + q^{r+s-1} &= \deg\left(\mathcal{M}_{r-1,d}\right) + d + q^{r+s-1} - s \\
&= \left(\deg\left(\mathcal{P}^{\Delta_s}\right) + d\right) q^r.
\end{aligned}$$

That's what we wanted. •

As we pointed out in Chapter 1 the recursion formula for the elements $\mathcal{M}_{r,d}$ leads to the following explicit description

$$\mathcal{M}_{r,d} = \mathcal{P}^{d_{r-1}-1} \cdots \mathcal{P}^{d_0-1}\mathcal{P}^0.$$

Finally:

LEMMA A.1.3: *The degree of the new elements $\mathcal{M}_{r,d}$ is given by*

$$\deg\left(\mathcal{M}_{r,d}\right) = q^r d_0 - q^r - d_0 + 1.$$

Moreover, there is the relation

$$d_r = q^r d_0 - q^r + 1.$$

PROOF: Again we induct on r. If $r = 0$ there is nothing to show. Let $r > 0$. Then

$$\begin{aligned}\deg\left(\mathcal{M}_{r,d}\right) &= \deg\left(\mathcal{M}_{r-1,d}\right) + \deg\left(\mathcal{P}^{d_{r-1}-1}\right)\\ &= q^{r-1}d_0 - q^{r-1} - d_0 + 1 + (d_{r-1} - 1)(q-1)\\ &= q^{r-1}d_0 - q^{r-1} - d_0 + 1 + (q^{r-1}d_0 - q^{r-1})(q-1)\\ &= q^{r-1}d_0 - q^{r-1} - d_0 + 1 + q^r d_0 - q^r - q^{r-1}d_0 + q^{r-1}\\ &= q^r d_0 - q^r - d_0 + 1,\end{aligned}$$

where we made use of the induction hypothesis for both formulae. Finally,

$$d_r = \deg\left(\mathcal{M}_{r,d}\right) + d_0$$

leads to the desired second formula. •

A.2 The Action of the Steenrod Algebra on the Dickson Algebra

In this section we want to show that the formulae for the action of the mod p Steenrod algebra on the Dickson algebra given in Section 3 of [40] and Section II of [45][1], have direct analogues over arbitrary finite fields. Let

$$\mathcal{D}^*(n) = \mathbb{F}[\mathbf{d}_{n,0}, \ldots, \mathbf{d}_{n,n-1}] \subseteq \mathbb{F}[V]^{\mathrm{GL}(n,\,\mathbb{F})}$$

be the Dickson algebra of Krull dimension n.

Recall from [36] Theorem 8.1.6 that the Dickson classes are given by the Stong-Tamagawa-formulae

$$\mathbf{d}_{n,i} = \sum_{W^* \leq V^*,\ \dim(W^*)=i} \left(\prod_{v \notin W^*} v\right).$$

We could equally have said that the Dickson classes occur as coefficients in the following polynomial

$$f(X) = \prod_{v \in V^*}(X + v) = X^{q^n} + \sum_{i=0}^{n-1} \mathbf{d}_{n,i} X^{q^i}.$$

To simplify the notation write

$$\mathbf{d}_{n,n} := 1 \quad \text{and} \quad \mathbf{d}_{n,i} := 0\ \forall\ i < 0.$$

[1] Note carefully that in [45] some formulae are not quite correct.

Then using the product form of $f(X)$, we get by the Cartan formulae,

$$\begin{aligned}
\mathcal{P}^k(f(X)) &= \mathcal{P}^k\left(\prod_{v\in V^*}(X+v)\right)\\
&= \sum_{(i_1,\dots,i_k)}\left(\mathcal{P}^1(X+v_{i_1})\cdots\mathcal{P}^1(X+v_{i_k})\right)\left(\prod_{V^*\backslash\{v_{i_1},\dots,v_{i_k}\}}(X+v)\right)\\
&= \sum_{(i_1,\dots,i_k)}\left((X^q+v_{i_1}^q)\cdots(X^q+v_{i_k}^q)\right)\left(\prod_{V^*\backslash\{v_{i_1},\dots,v_{i_k}\}}(X+v)\right)\\
&= f(X)\left(\sum_{(i_1,\dots,i_k)}(X+v_{i_1})^{q-1}\cdots(X+v_{i_k})^{q-1}\right),
\end{aligned}$$

i.e., $f(X)$ divides every Steenrod power of itself. On the other hand we can use the sum form of $f(X)$. Then we get

$$\mathcal{P}^k\left(f(X)\right) = \mathcal{P}^k\left(X^{q^n}+\sum_{i=0}^{n-1}\mathbf{d}_{n,i}X^{q^i}\right)$$

$$= \begin{cases} 0 & \text{if } k > q^n\\ X^{q^{n+1}}+\sum_{i=0}^{n-1}\left(\mathcal{P}^{q^n}(\mathbf{d}_{n,i})X^{q^i}+\mathcal{P}^{q^n-q^i}(\mathbf{d}_{n,i})\mathcal{P}^{q^i}(X^{q^i})\right) & \text{if } k = q^n\\ \sum_{i=0}^{n-1}\left(\mathcal{P}^k(\mathbf{d}_{n,i})X^{q^i}+\mathcal{P}^{k-q^i}(\mathbf{d}_{n,i})\mathcal{P}^{q^i}(X^{q^i})\right) & \text{if } 1\le k< q^n\\ f(X) & \text{if } k=0 \end{cases}$$

$$= \begin{cases} 0 & \text{if } k > q^n\\ X^{q^{n+1}}+\sum_{i=0}^{n-1}\mathcal{P}^{q^n-q^i}(\mathbf{d}_{n,i})\mathcal{P}^{q^i}(X^{q^i}) & \text{if } k = q^n\\ \sum_{i=0}^{n-1}\left(\mathcal{P}^k(\mathbf{d}_{n,i})X^{q^i}+\mathcal{P}^{k-q^i}(\mathbf{d}_{n,i})X^{q^{i+1}}\right) & \text{if } 1\le k< q^n\\ f(X) & \text{if } k=0 \end{cases}$$

$$= \begin{cases} 0 & \text{if } k > q^n\\ X^{q^{n+1}}+\sum_{i=0}^{n-1}(\mathbf{d}_{n,i})^q X^{q^{i+1}} & \text{if } k = q^n\\ \sum_{i=0}^{n}\left(\mathcal{P}^k(\mathbf{d}_{n,i})+\mathcal{P}^{k-q^{i-1}}(\mathbf{d}_{n,i-1})\right)X^{q^i} & \text{if } 1\le k< q^n\\ f(X) & \text{if } k=0. \end{cases}$$

Combining the two formulae and comparing the highest coefficients gives for $1 \le k < q^n$

$$\mathcal{P}^k\left(f(X)\right) = f(X)\mathcal{P}^{k-q^{n-1}}(\mathbf{d}_{n,n-1}).$$

For degree reasons

$$\mathcal{P}^k(\mathbf{d}_{n,i}) = 0 \quad \text{if } k \geq q^n,$$

hence we have

$$\mathcal{P}^k(\mathbf{d}_{n,i}) = \begin{cases} 0 & \text{if } k \geq q^n \\ \mathcal{P}^{k-q^{n-1}}(\mathbf{d}_{n,n-1})\mathbf{d}_{n,i} - \mathcal{P}^{k-q^{i-1}}\mathbf{d}_{n,i-1} & \text{if } 1 \leq k < q^n \\ \mathbf{d}_{n,i} & \text{if } k = 0. \end{cases}$$

This corrects Corollary 2.3 a) in [45] and extends it from $\mathbb{F}_p$ to $\mathbb{F}_q$.

Analogously, one proves the formulae for $\mathcal{P}^{\Delta_k}(\mathbf{d}_{n,i})$. First we use the product form for $f(X)$ and get

$$\begin{aligned} \mathcal{P}^{\Delta_k}\left(f(X)\right) &= \sum_{v_1 \in V^*} \mathcal{P}^{\Delta_k}(X + v_1) \prod_{v \in V^* \setminus \{v_1\}} (X + v) \\ &= \sum_{v_1 \in V^*} (X^{q^k} + v_1^{q^k}) \prod_{v \in V^* \setminus \{v_1\}} (X + v) \\ &= f(X) \sum_{v_1 \in V^*} (X + v_1)^{q^k - 1}. \end{aligned}$$

Here we made use of the derivation property of the $\mathcal{P}^{\Delta_k}$'s and how they act on linear forms. Again

$$f(X) \mid \mathcal{P}^{\Delta_k}\left(f(X)\right).$$

Using the sum formula for $f(X)$ we get

$$\mathcal{P}^{\Delta_k}\left(f(X)\right) = \mathbf{d}_{n,0}X^{q^k} + \sum_{i=0}^{n-1} \left(\mathcal{P}^{\Delta_k}\left(\mathbf{d}_{n,i}\right) X^{q^i}\right).$$

Once again we compare the coefficients in the two formulae and use the defining recursion formulae for the $\mathcal{P}^{\Delta}$'s, to get

$$\mathcal{P}^{\Delta_k}(f(X)) = \begin{cases} 0 & \text{for } k < n \\ \mathbf{d}_{n,0}f(X) & \text{for } k = n \\ \mathcal{P}^{q^{k-1}}\mathcal{P}^{\Delta_{k-1}}\left(f(X)\right) & \text{for } k > n, \end{cases}$$

where we made use of what we have proven above. Recursively we recover the formulae of [40] Section 3, see also [36] §10.6, or [45] Corollary 2.3 b, (in the latter again the signs are not correct) extended to $\mathbb{F}_q$

$$\mathcal{P}^{\Delta_k}(\mathbf{d}_{n,i}) = \begin{cases} 0 & \text{if } 0 \leq k < n \text{ and } k \neq i \\ -\mathbf{d}_{n,0} & \text{if } 0 \leq k < n \text{ and } k = i \\ \mathbf{d}_{n,0}\mathbf{d}_{n,i} & \text{if } k = n \\ \mathcal{P}^{q^{k-1}}\mathcal{P}^{\Delta_{k-1}}\left(\mathbf{d}_{n,i}\right) & \text{if } k > n \end{cases}.$$

We collect these results in a proposition.

PROPOSITION A.2.1: *With the above notation we have*

(1)

$$\mathcal{P}^k(\mathbf{d}_{n,i}) = \begin{cases} 0 & \text{if } k \geq q^n \\ \mathcal{P}^{k-q^{n-1}}(\mathbf{d}_{n,n-1})\mathbf{d}_{n,i} - \mathcal{P}^{k-q^{i-1}}(\mathbf{d}_{n,i-1}) & \text{if } 1 \leq k < q^n \\ \mathbf{d}_{n,i} & \text{if } k = 0. \end{cases}$$

(2)

$$\mathcal{P}^{\Delta_k}(\mathbf{d}_{n,i}) = \begin{cases} 0 & \text{if } 0 \leq k < n \text{ and } k \neq i \\ -\mathbf{d}_{n,0} & \text{if } 0 \leq k < n \text{ and } k = i \\ \mathbf{d}_{n,0}\mathbf{d}_{n,i} & \text{if } k = n \\ \mathcal{P}^{q^{k-1}}\mathcal{P}^{\Delta_{k-1}}\left(\mathbf{d}_{n,i}\right) & \text{if } k > n. \end{cases}$$

•

The following extends the analogous formulae of §3 in [40], see also Corollary 2.4 of [45].

COROLLARY A.2.2: *With the above notation we have*

$$\mathcal{P}^{q^k}(\mathbf{d}_{n,i}) = \begin{cases} -\mathbf{d}_{n,i-1} & \text{for } k = i-1 \geq 0 \\ \mathbf{d}_{n,i}\mathbf{d}_{n,n-1} & \text{for } k = n-1 \geq 0 \\ 0 & \text{otherwise} \end{cases}$$

for all $i = 0, \ldots, n-1$.

PROOF: This formula can be read off directly from part (1) of the preceding proposition. •

A.3 The Fractal Property of the Dickson Algebra

For the sake of completeness we will prove the fractal property of the Dickson algebra in two versions, compare also proof of Theorem 8.1.6 in [36].

PROPOSITION A.3.1 (Fractal Property, Version 1): *Consider the projection*

$$\mathrm{pr} : V^* := \mathrm{Span}_{\mathbb{F}}\{z_1, \ldots, z_n\} \longrightarrow W^* := \mathrm{Span}_{\mathbb{F}}\{z_2, \ldots, z_n\}.$$

Then the induced map on the polynomial algebras $\mathbb{F}[V] \xrightarrow{\mathrm{pr}} \mathbb{F}[W]$ *maps the Dickson algebra* $\mathcal{D}^*(n)$ *onto* $\mathcal{D}^*(n-1)^q$.

PROOF: We use the Stong-Tamagawa formula given in the preceding section:

$$\begin{aligned} \mathrm{pr}(\mathbf{d}_{n,i}) &= \mathrm{pr}\left(\sum_{V'^* \leq V^*,\ \dim(V'^*)=i} \left(\prod_{v \notin V'^*} v\right)\right) \\ &= \mathrm{pr}\left(\sum_{V'^* \leq V^*,\ \dim(V'^*)=i} \left(\prod_{v=\bar{v}+\lambda z_1 \notin V'^*,\ \lambda \in \mathbb{F}} (\bar{v} + \lambda z_1)\right)\right) \end{aligned}$$

$$= \sum_{V'^* \leq V^*,\ \dim(V'^*)=i, z_1 \in V'^*} \left(\prod_{v=\bar{v}+\lambda z_1 \notin V'^*,\ \lambda \in \mathbb{F}} (\bar{v}) \right)$$

$$= \sum_{V'^* \leq V^*,\ \dim(V'^*)=i, z_1 \in V'^*} \left(\prod_{v=\bar{v} \notin V'^*} \bar{v}^q \right)$$

$$= \sum_{W'^* \leq W^*,\ \dim(W'^*)=i-1} \left(\prod_{v=\bar{v} \notin W'^*} \bar{v}^q \right)$$

$$= \mathbf{d}^q_{n-1,i-1}$$

giving what we wanted. •

PROPOSITION A.3.2 (Fractal Property, Version 2): *We have the following isomorphism in the category of unstable algebras*

$$\mathcal{D}^*(n)/(\mathbf{d}_{n,0}) \cong \mathcal{D}^*(n-1)^q.$$

PROOF: Let $\mathrm{pr} : \mathcal{D}^*(n) \longrightarrow \mathcal{D}^*(n)/(\mathbf{d}_{n,0})$ be the canonical projection and denote by

$$\overline{\mathbf{d}_{n,i}} = \mathrm{pr}(\mathbf{d}_{n,i}) \ \forall\ i = 0, \dots, n-1$$

the images of the Dickson classes. Define a map

$$\mathcal{D}^*(n)/(\mathbf{d}_{n,0}) = \mathbb{F}[\overline{\mathbf{d}_{n,1}}, \dots, \overline{\mathbf{d}_{n,n-1}}] \xrightarrow{\varphi} \mathcal{D}^*(n-1)^q$$

by

$$\varphi\left(\overline{\mathbf{d}_{n,i}}\right) := \mathbf{d}^q_{n-1,i-1}.$$

We have to show that this map commutes with Steenrod powers. For $\mathcal{P}^k$ with $k = 0$ or $k \geq q^n$ nothing needs to be shown. So we induct on k, and assume in the following, that $1 \leq k < q^n$. Recalling our convention, from Section 1.1, that $\mathcal{P}^i \equiv 0$ if $i \notin \mathbb{N}_0$, we have

$$\begin{aligned}
&\varphi\left(\mathcal{P}^k\left(\overline{\mathbf{d}_{n,i}}\right)\right)\\
&\overset{(1)}{=} \varphi\left(-\mathcal{P}^{k-q^{i-1}}\left(\overline{\mathbf{d}_{n,i-1}}\right)\right) + \varphi\left(\mathcal{P}^{k-q^{n-1}}\left(\overline{\mathbf{d}_{n,n-1}}\right)\overline{\mathbf{d}_{n,i}}\right)\\
&\overset{(2)}{=} -\mathcal{P}^{k-q^{i-1}}\left(\varphi\left(\overline{\mathbf{d}_{n,i-1}}\right)\right) + \mathcal{P}^{k-q^{n-1}}\left(\varphi\left(\overline{\mathbf{d}_{n,n-1}}\right)\right)\varphi\left(\overline{\mathbf{d}_{n,i}}\right)\\
&\overset{(3)}{=} -\mathcal{P}^{k-q^{i-1}}\left(\mathbf{d}^q_{n-1,i-2}\right) + \mathcal{P}^{k-q^{n-1}}\left(\mathbf{d}^q_{n-1,n-2}\right)\mathbf{d}^q_{n-1,i-1}\\
&\overset{(4)}{=} -\left(\mathcal{P}^{\frac{k-q^{i-1}}{q}}\left(\mathbf{d}_{n-1,i-2}\right)\right)^q + \left(\mathcal{P}^{\frac{k-q^{n-1}}{q}}\left(\mathbf{d}_{n-1,n-2}\right)\right)^q \mathbf{d}^q_{n-1,i-1}\\
&\overset{(5)}{=} \left(-\mathcal{P}^{\frac{k-q^{i-1}}{q}}\left(\mathbf{d}_{n-1,i-2}\right) + \mathcal{P}^{\frac{k-q^{n-1}}{q}}\left(\mathbf{d}_{n-1,n-2}\right)\mathbf{d}_{n-1,i-1}\right)^q
\end{aligned}$$

$$\stackrel{(6)}{=} \left(\mathcal{P}^{\frac{k}{q}}\left(\mathbf{d}_{n-1,i-1}\right)\right)^{q}$$
$$\stackrel{(7)}{=} \mathcal{P}^{k}\left(\mathbf{d}_{n-1,i-1}^{q}\right)$$
$$\stackrel{(8)}{=} \mathcal{P}^{k}\left(\varphi\left(\overline{\mathbf{d}_{n,i}}\right)\right),$$

where

(1) follows from the fact that the projection commutes with the Steenrod action and the explicit formulae given in the preceding section for the Steenrod powers of the Dickson classes,
(2) follows from induction,
(3) and (8) by definition of φ,
(4) and (7) by the Cartan formulae,
(5) from the additivity of the Frobenius map in characteristic p, and
(6) again from the formulae of the preceding section.

•

A.4 The Generalized Jacobian

A reference for the following result is hard to find in the literature. Therefore it is included here. However, this is just a careful extension of the proof given in Lemma 5.6.1 in [36] for the classical situation.

THEOREM A.4.1: *Let* H^* *be an unstable algebra over* $\mathcal{P}^*$, $h_0, \ldots, h_m \in \mathrm{H}^*$. *If the determinant of the generalized Jacobian matrix*

$$\begin{bmatrix} \mathcal{P}^{\Delta_0}(h_0) & \cdots & \mathcal{P}^{\Delta_0}(h_m) \\ \vdots & \ddots & \vdots \\ \mathcal{P}^{\Delta_m}(h_0) & \cdots & \mathcal{P}^{\Delta_m}(h_m) \end{bmatrix}$$

is neither zero nor a zero divisor, then the elements $h_0, \ldots, h_m$ *are algebraically independent.*

PROOF: If the generalized Jacobian is not a zero divisor (or zero) in H^*, then this remains true if we consider it in the quotient $\mathrm{H}^*/\mathcal{N}il(\mathrm{H}^*)$. Suppose to the contrary that $h_0, \ldots, h_m$ were algebraically dependent in H^*. If they are also algebraically dependent in $\mathrm{H}^*/\mathcal{N}il(\mathrm{H}^*)$ then choose a polynomial $f(X_0, \ldots, X_m) \in \mathbb{F}[X_0, \ldots, X_m]$ of minimal degree such that

$$f(h_0, \ldots, h_m) = 0.$$

Then certainly $\mathcal{P}^{\Delta_j}\left(f(h_0, \ldots, h_m)\right) = 0$ for any $j \geq 0$. Write

$$f(h_0, \ldots, h_m) = \sum_{j_0, \ldots, j_m} \lambda_J h_0^{j_0} \cdots h_m^{j_m},$$

for some coefficients $\lambda_J \in \mathbb{F}^\times$ with indices $J = J(j_0, \ldots, j_m)$. Then we have that

$$\mathcal{P}^{\Delta_i}\left(f(h_0, \ldots, h_m)\right) = \sum_{j_0,\ldots,j_m} \sum_{k=0}^{m} j_k \lambda_J h_0^{j_0} \cdots \mathcal{P}^{\Delta_i}(h_k) h_k^{j_k-1} \cdots h_m^{j_m} = 0.$$

Hence

$$\begin{bmatrix} 0 \\ \vdots \\ 0 \end{bmatrix} = \begin{bmatrix} \mathcal{P}^{\Delta_0}\left(f(h_0, \ldots, h_m)\right) \\ \vdots \\ \mathcal{P}^{\Delta_m}\left(f(h_0, \ldots, h_m)\right) \end{bmatrix}$$

$$= \begin{bmatrix} \mathcal{P}^{\Delta_0}(h_0) & \cdots & \mathcal{P}^{\Delta_0}(h_m) \\ \vdots & \ddots & \vdots \\ \mathcal{P}^{\Delta_m}(h_0) & \cdots & \mathcal{P}^{\Delta_m}(h_m) \end{bmatrix} \begin{bmatrix} \sum\limits_{j_0>0,\ j_1,\ldots,j_m} j_0 \lambda_J h_0^{j_0-1} h_1^{j_1} \cdots h_m^{j_m} \\ \vdots \\ \sum\limits_{j_m>0,\ j_0,\ldots,j_m} j_m \lambda_J h_0^{j_0} \cdots h_{m-1}^{j_{m-1}} h_m^{j_m-1} \end{bmatrix}.$$

By assumption the determinant of the generalized Jacobian matrix is neither zero nor a zero divisor. Therefore the vector

$$\begin{bmatrix} \sum\limits_{j_0>0,\ j_1,\ldots,j_m} j_0 \lambda_J h_0^{j_0-1} h_1^{j_1} \cdots h_m^{j_m} \\ \vdots \\ \sum\limits_{j_m>0,\ j_0,\ldots,j_m} j_m \lambda_J h_0^{j_0} \cdots h_{m-1}^{j_{m-1}} h_m^{j_m-1} \end{bmatrix}$$

must be the zero vector. This hands us polynomials

$$f_i(X_0, \ldots, X_m) := \sum_{j_i>0, j_0,\ldots,j_m} j_i \lambda_J X_0^{j_0} \cdots X_i^{j_i-1} \cdots X_m^{j_m}$$

for $i = 0, \ldots, m$, of degree less than $f(X_0, \ldots, X_m)$ which evaluates to zero on the elements $h_0, \ldots, h_m$. Therefore these must be the zero polynomials, i.e, the coefficients

$$j_0, \ldots, j_m$$

are divisible by the characteristic of the ground field $\mathbb{F}$. This in turn implies that

$$\tilde{f}(h_0, \ldots, h_m) = \sum_{j_0,\ldots,j_m} \lambda_J h_0^{\frac{j_0}{p}} \cdots h_m^{\frac{j_m}{p}}$$

is a non-zero polynomial of degree less than $f(X_0, \ldots, X_m)$ evaluating to zero on the elements $h_0, \ldots, h_m$. This is a contradiction. Hence $h_0, \ldots, h_m$ are algebraically independent in $\mathrm{H}^*/\mathcal{N}il(\mathrm{H}^*)$, hence they are so in H^*. •

Remark: Note that this proof would also work for an arbitrary set of derivations, i.e., we have not used the particular form of the $\mathcal{P}^{\Delta_i}$'s.

References

[1] J. F. Adams and C. W. Wilkerson: *Finite H-Spaces and Algebras over the Steenrod Algebra*, Annals of Math. 111 (1980), 95-143.

[2] J. F. Adams and C. W. Wilkerson: *Finite H-Spaces and Algebras over the Steenrod Algebra; a correction*, Annals of Math. 113 (1981), 621-622.

[3] M. F. Atiyah and I. G. Macdonald: *Introduction to Commutative Algebra*, Addison-Wesley Publishing Company, Reading Massachusetts 1969.

[4] Stanislaw Balcerzyk and Tadeusz Józefiak: *Commutative Noetherian and Krull Rings*, Ellis Horwood Limited, Chichester 1989.

[5] Stanislaw Balcerzyk and Tadeusz Józefiak: *Commutative Rings. Dimension, Multiplicity and Homological Methods*, Ellis Horwood Limited, Chichester 1989.

[6] Dorra Bourguiba and Saïd Zarati: *Depth and Steenrod Operations*, Inventiones Mathematicae 128 (1997), 589-602.

[7] S. R. Bullett and I. G. Macdonald: *On the Adem Relations*, Topology 21 (1982), 329-332.

[8] Henri Cartan: *Sur l'Itération des Opérations de Steenrod*, Commentarii Mathematici Helvetici 29 (1955), 40-58.

[9] Leonard E. Dickson: *A Fundamental System of Invariants of the General Modular Linear Group with a Solution of the Form Problem*, Transactions of the AMS 12 (1911), 175-192.

[10] Jeanne Duflot: *Depth and Equivariant Cohomology*, Commentarii Mathematici Helvetici 56 (1981), 627-637.

[11] Jeanne Duflot: *Lots of Hopf Algebras*, Journal of Algebra 204 (1998), 69-94.

[12] Jeanne Duflot: *Private Communication*, 1999.

[13] Jeanne Duflot, Nicholas J. Kuhn and Mark Winstead: *A Classification of Polynomial Algebras as Modules over the Steenrod Algebra*, Commentarii Mathematici Helvetici 68 (1993), 622-632.

[14] William G. Dwyer and Clarence W. Wilkerson: *Smith Theory Revisited*, Annals of Math. 127 (1988), 191-198.

[15] Paul M. Eakin, Jr.: *The Converse to a Well Known Theorem on Noetherian Rings*, Math. Annalen 177 (1968), 278-282.

[16] David Eisenbud: *Commutative Algebra*, Graduate Texts in Mathematics 150, Springer Verlag, New York 1995.

[17] Nathan Jacobson: *Lectures in Abstract Algebra. Volume III*, Graduate Texts in Mathematics 32, Springer Verlag, New York-Heidelberg-Berlin 3rd corr. printing 1980.

[18] Richard M. Kane: *The Homology of Hopf Spaces*, North-Holland Mathematical Library 40, North-Holland, Amsterdam-New York-Oxford-Tokyo 1988.

[19] Siu-Por Lam: *Unstable Algebras over the Steenrod Algebra and Cohomology of Classifying Spaces*, Proceedings of the Conference on Algebraic Topology held in Aarhus 1982, Springer Verlag, Heidelberg-Berlin 1983.

[20] Peter S. Landweber: *Dickson Invariants and Prime Ideals Invariant under Steenrod Operations*, seminar talk, Princeton 1984.

[21] Peter S. Landweber: *Dickson Invariants and Steenrod Operations on Cohomology Rings*, talk at the AMS Summer Conference on Algebraic Topology held at Minneapolis 1984.

[22] Peter S. Landweber and Robert S. Stong: *The Depth of Rings of Invariants over Finite Fields*, in: Proceedings of the New York Number Theory seminar 1984, Lecture Notes in Mathematics 1240, Springer Verlag, New York 1987.

[23] J. Lannes and S. Zarati: *Théorie de Smith algébrique et classification des* H^*V – $\mathcal{U}$*-injectifs*, Bull. Soc. Math. de France 123 (1995), 189 -224.

[24] W. S. Massey and F. P. Peterson: *The Cohomology Structure of Certain Fibre Spaces I*, Topology 4 (1965), 47-65.

[25] John W. Milnor: *The Steenrod Algebra and its Dual*, Annals of Math. 67 (1958), 150-171.

[26] Mara D. Neusel: *Integral Extensions of Unstable Algebras over the Steenrod Algebra*, Forum Mathematicum, to appear.

[27] Mara D. Neusel: *Localizations over the Steenrod Algebra. The lost Chapter*, preprint, Wyoming 1998.

[28] Mara D. Neusel and Larry Smith: *Thom Classes and the Landweber-Stong-Conjecture*, unpublished manuscript, Oberseminar Göttingen 1996.

[29] Mara D. Neusel and Larry Smith: *The Lasker-Noether Theorem for* $\mathcal{P}^*$*-Invariant Ideals*, Forum Mathematicum 10 (1998), 1-18.

[30] D. L. Rector: *Noetherian Cohomology Rings and Finite Loop Spaces with Torsion*, Journal of Pure and Applied Algebra 32 (1984), 191-217.

[31] Leslie G. Roberts: *Private Communication*, 1997-1998.

[32] Jean-Pierre Serre: *Cohomologie modulo 2 des Complexes d'Eilenberg-Maclane*, Commentarii Mathematici Helvetici 27 (1953), 198-232.

[33] Jean-Pierre Serre: *Sur la Dimension Cohomologique des Groupes Profinis*, Topology 3 (1965), 413-420.

[34] Larry Smith: *Modular Representations of q-Groups with Regular Rings of Invariants*, preprint, Stockholm University 1996.

[35] Larry Smith: *$\mathcal{P}^*$-Invariant Ideals in Rings of Invariants*, Forum Mathematicum 8 (1996), 319-342.

[36] Larry Smith: *Polynomial Invariants of Finite Groups*, 2nd corrected printing, A.K. Peters Ltd., Wellesley Mass. 1997.

[37] Larry Smith: *Polynomial Invariants of Finite Groups. A Survey of Recent Developments*, Bulletin of the AMS 34 (1997), 211-250.

[38] Larry Smith: *Private Communication*, 1997-1998.

[39] Larry Smith and Robert E. Stong: *On the Invariant Theory of Finite Groups: Orbit Polynomials and Splitting Principles*, Journal of Algebra 110 (1987), 134-157.

[40] Larry Smith and R. M. Switzer: *Polynomial algebras over the Steenrod algebra. Variations on a Theorem of Adams and Wilkerson*, Proceedings of the Edinburgh Mathematical Society 27 (1984), 11-19.

[41] N. E. Steenrod and D. B. A. Epstein: *Cohomology Operations*, Annals of Mathematics Studies 50, Princeton University Press, Princeton NJ 1962.

[42] Charles A. Weibel: *An Introduction to Homological Algebra*, Cambridge Studies in Advanced Mathematics 38, Cambridge University Press, Cambridge 1994.

[43] Clarence W. Wilkerson: *Classifying Spaces, Steenrod Operations and Algebraic Closure*, Topology 16 (1977), 227-237.

[44] Clarence W. Wilkerson: *Integral Closure of Unstable Steenrod Algebra Actions*, Journal of Pure and Applied Algebra 13 (1978), 49-55.

[45] Clarence W. Wilkerson: *A Primer on the Dickson Invariants*, Proceedings of the Northwestern Homotopy Theory Conference, Evanston, Contemporary Mathematics 19, 1983.

[46] Clarence W. Wilkerson: *Rings of Invariants and Inseparable Forms of Algebras over the Steenrod Algebra*, preprint, Purdue University 19??.

[47] Wu Wen-Tsün: *Sur les Puissance de Steenrod*, Colloque de Topologie de Strasbourg, 1951.

[48] O. Zariski and P. Samuel: *Commutative algebra. Volume I*, D. van Nostrand Company, Inc., Princeton NJ 1958.

[49] O. Zariski and P. Samuel: *Commutative algebra. Volume II*, D. van Nostrand Company, Inc., Princeton NJ 1960.

Font and Typesetting Information

For the typesetting we used $\mathcal{LS}$TEX with Postscript[1] fonts: New Century Schoolbook roman was used for the running text, *italic* for emphasis and **bold** for notions to be defined. The statements of theorems, propositions, etc. are in *oblique roman* created by Postscript hackery. For fields, such as $\mathbb{F}$ and $\mathbb{K}^*$, we used an outline version of Charter BT-Bold from Bitstream. Hermann Zapf's *Chancery Medium Italic* is used for, e.g., nil radicals $\mathcal{Nil}$, the Steenrod algebra $\mathcal{P}^*$ and the Dickson algebra $\mathcal{D}^*(n)$. We used **Helvetica Bold** for, e.g., the Dickson classes $\mathbf{d}_{n,0}$, as well as for linear transformations and matrices $\mathbf{M}$. To denote the giant Steenrod operation P we used URW Dom Casual. The Steenrod reduced powers $\mathcal{P}^i$ are from Linoscript from LinoType Hell GmbH, and the greek letters, γ, from the library of characters designed by Dr. A. V. Hershey of the US Bureau of Standards. The german letters for ideals, such as $\mathfrak{p}$, are from Alte Schwabacher from URW++. The mathematical symbols come mostly from the font Shapes designed[2] by Larry Smith and the Adobe Math Symbol Font. Finally, the wisecracks appear in the font Dijkstra.

[1] Postscript is a copyright of Adobe Systems Incorporated.

[2] Available from URW++ Design & Development GmbH, Hamburg, Germany, http://www.urwpp.de.

Mara D. Neusel

Mittelweg 3

D37133 Friedland
Germany
mdn@sunrise.uni-math.gwdg.de

current address

Department of Mathematics

Yale University
10 Hillhouse Avenue
P. O. Box 208283
New Haven CT 06520-8283
USA

Editorial Information

To be published in the *Memoirs*, a paper must be correct, new, nontrivial, and significant. Further, it must be well written and of interest to a substantial number of mathematicians. Piecemeal results, such as an inconclusive step toward an unproved major theorem or a minor variation on a known result, are in general not acceptable for publication. *Transactions* Editors shall solicit and encourage publication of worthy papers. Papers appearing in *Memoirs* are generally longer than those appearing in *Transactions* with which it shares an editorial committee.

As of March 31, 2000, the backlog for this journal was approximately 7 volumes. This estimate is the result of dividing the number of manuscripts for this journal in the Providence office that have not yet gone to the printer on the above date by the average number of monographs per volume over the previous twelve months, reduced by the number of issues published in four months (the time necessary for preparing an issue for the printer). (There are 6 volumes per year, each containing at least 4 numbers.)

A Copyright Transfer Agreement is required before a paper will be published in this journal. By submitting a paper to this journal, authors certify that the manuscript has not been submitted to nor is it under consideration for publication by another journal, conference proceedings, or similar publication.

Information for Authors and Editors

Memoirs are printed by photo-offset from camera copy fully prepared by the author. This means that the finished book will look exactly like the copy submitted.

The paper must contain a *descriptive title* and an *abstract* that summarizes the article in language suitable for workers in the general field (algebra, analysis, etc.). The *descriptive title* should be short, but informative; useless or vague phrases such as "some remarks about" or "concerning" should be avoided. The *abstract* should be at least one complete sentence, and at most 300 words. Included with the footnotes to the paper, there should be the 2000 *Mathematics Subject Classification* representing the primary and secondary subjects of the article. This may be followed by a list of *key words and phrases* describing the subject matter of the article and taken from it. A list of the numbers may be found in the annual index of *Mathematical Reviews*, published with the December issue starting in 1990, as well as from the electronic service e-MATH [**telnet e-MATH.ams.org** (or **telnet 130.44.1.100**). Login and password are **e-math**]. For journal abbreviations used in bibliographies, see the list of serials on the web at `http://www.ams.org/msnhtml/serials-list/annser_frames.html`. When the manuscript is submitted, authors should supply the editor with electronic addresses if available. These will be printed after the postal address at the end of each article.

Electronically prepared papers. The AMS encourages submission of electronically prepared papers in $\mathcal{A}\mathcal{M}\mathcal{S}$-TeX or $\mathcal{A}\mathcal{M}\mathcal{S}$-LaTeX. The Society has prepared author packages for each AMS publication. Author packages include instructions for preparing electronic papers, the *AMS Author Handbook*, samples, and a style file that generates the particular design specifications of that publication series for both $\mathcal{A}\mathcal{M}\mathcal{S}$-TeX and $\mathcal{A}\mathcal{M}\mathcal{S}$-LaTeX.

Authors with FTP access may retrieve an author package from the Society's Internet node `e-MATH.ams.org` (130.44.1.100). For those without FTP access, the author package can be obtained free of charge by sending e-mail to pub@ams.org (Internet) or from the Publication Division, American Mathematical Society, P.O. Box 6248, Providence, RI 02940-6248. When requesting an author package, please specify $\mathcal{AMS}$-TeX or $\mathcal{AMS}$-LaTeX, Macintosh or IBM (3.5) format, and the publication in which your paper will appear. Please be sure to include your complete mailing address.

Submission of electronic files. At the time of submission, the source file(s) should be sent to the Providence office (this includes any TeX source file, any graphics files, and the DVI or PostScript file).

Before sending the source file, be sure you have proofread your paper carefully. The files you send must be the EXACT files used to generate the proof copy that was accepted for publication. For all publications, authors are required to send a printed copy of their paper, which exactly matches the copy approved for publication, along with any graphics that will appear in the paper.

TeX files may be submitted by email, FTP, or on diskette. The DVI file(s) and PostScript files should be submitted only by FTP or on diskette unless they are encoded properly to submit through e-mail. (DVI files are binary and PostScript files tend to be very large.)

Files sent by electronic mail should be addressed to the Internet address pub-submit@ams.org. The subject line of the message should include the publication code to identify it as a Memoir. TeX source files, DVI files, and PostScript files can be transferred over the Internet by FTP to the Internet node e-math.ams.org (130.44.1.100).

Electronic graphics. Figures may be submitted to the AMS in an electronic format. The AMS recommends that graphics created electronically be saved in Encapsulated PostScript (EPS) format. This includes graphics originated via a graphics application as well as scanned photographs or other computer-generated images.

If the graphics package used does not support EPS output, the graphics file should be saved in one of the standard graphics formats—such as TIFF, PICT, GIF, etc.—rather than in an application-dependent format. Graphics files submitted in an application-dependent format are not likely to be used. No matter what method was used to produce the graphic, it is necessary to provide a paper copy to the AMS.

Authors using graphics packages for the creation of electronic art should also avoid the use of any lines thinner than 0.5 points in width. Many graphics packages allow the user to specify a "hairline" for a very thin line. Hairlines often look acceptable when proofed on a typical laser printer. However, when produced on a high-resolution laser imagesetter, hairlines become nearly invisible and will be lost entirely in the final printing process.

Screens should be set to values between 15% and 85%. Screens which fall outside of this range are too light or too dark to print correctly.

Any inquiries concerning a paper that has been accepted for publication should be sent directly to the Editorial Department, American Mathematical Society, P. O. Box 6248, Providence, RI 02940-6248.

Editors

This journal is designed particularly for long research papers (and groups of cognate papers) in pure and applied mathematics. Papers intended for publication in the *Memoirs* should be addressed to one of the following editors. In principle the Memoirs welcomes electronic submissions, and some of the editors, those whose names appear below with an asterisk (*), have indicated that they prefer them. However, editors reserve the right to request hard copies after papers have been submitted electronically. Authors are advised to make preliminary e-mail inquiries to editors about whether they are likely to be able to handle submissions in a particular electronic form.

Selected Titles in This Series

(Continued from the front of this publication)

659 **R. D. Nussbaum and S. M. Verduyn Lunel,** Generalizations of the Perron-Frobenius theorem for nonlinear maps, 1999

658 **Hasna Riahi,** Study of the critical points at infinity arising from the failure of the Palais-Smale condition for n-body type problems, 1999

657 **Richard F. Bass and Krzysztof Burdzy,** Cutting Brownian paths, 1999

656 **W. G. Bade, H. G. Dales, and Z. A. Lykova,** Algebraic and strong splittings of extensions of Banach algebras, 1999

655 **Yuval Z. Flicker,** Matching of orbital integrals on $GL(4)$ and $GSp(2)$, 1999

654 **Wancheng Sheng and Tong Zhang,** The Riemann problem for the transportation equations in gas dynamics, 1999

653 **L. C. Evans and W. Gangbo,** Differential equations methods for the Monge-Kantorovich mass transfer problem, 1999

652 **Arne Meurman and Mirko Primc,** Annihilating fields of standard modules of $\mathfrak{sl}(2,\mathbb{C})^{\sim}$ and combinatorial identities, 1999

651 **Lindsay N. Childs, Cornelius Greither, David J. Moss, Jim Sauerberg, and Karl Zimmermann,** Hopf algebras, polynomial formal groups, and Raynaud orders, 1998

650 **Ian M. Musson and Michel Van den Bergh,** Invariants under Tori of rings of differential operators and related topics, 1998

649 **Bernd Stellmacher and Franz Georg Timmesfeld,** Rank 3 amalgams, 1998

648 **Raúl E. Curto and Lawrence A. Fialkow,** Flat extensions of positive moment matrices: Recursively generated relations, 1998

647 **Wenxian Shen and Yingfei Yi,** Almost automorphic and almost periodic dynamics in skew-product semiflows, 1998

646 **Russell Johnson and Mahesh Nerurkar,** Controllability, stabilization, and the regulator problem for random differential systems, 1998

645 **Peter W. Bates, Kening Lu, and Chongchun Zeng,** Existence and persistence of invariant manifolds for semiflows in Banach space, 1998

644 **Michael David Weiner,** Bosonic construction of vertex operator para-algebras from symplectic affine Kac-Moody algebras, 1998

643 **Józef Dodziuk and Jay Jorgenson,** Spectral asymptotics on degenerating hyperbolic 3-manifolds, 1998

642 **Chu Wenchang,** Basic almost-poised hypergeometric series, 1998

641 **W. Bulla, F. Gesztesy, H. Holden, and G. Teschl,** Algebro-geometric quasi-periodic finite-gap solutions of the Toda and Kac-van Moerbeke hierarchies, 1998

640 **Xingde Dai and David R. Larson,** Wandering vectors for unitary systems and orthogonal wavelets, 1998

639 **Joan C. Artés, Robert E. Kooij, and Jaume Llibre,** Structurally stable quadratic vector fields, 1998

638 **Gunnar Fløystad,** Higher initial ideals of homogeneous ideals, 1998

637 **Thomáš Gedeon,** Cyclic feedback systems, 1998

636 **Ching-Chau Yu,** Nonlinear eigenvalues and analytic-hypoellipticity, 1998

635 **Magdy Assem,** On stability and endoscopic transfer of unipotent orbital integrals on p-adic symplectic groups, 1998

634 **Darrin D. Frey,** Conjugacy of Alt_5 and $\mathrm{SL}(2,5)$ subgroups of $\mathrm{E}_8(\mathbb{C})$, 1998

For a complete list of titles in this series, visit the AMS Bookstore at **www.ams.org/bookstore/**.

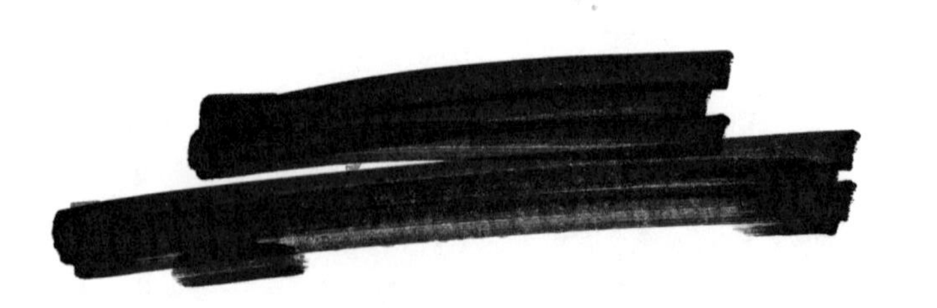